The V

Designing the Logic to Approximate Human Thinking

Chapman & Hall/CRC
Artificial Intelligence and Robotics Series

Series Editor
Roman Yampolskiy
University of Louisville
Louisville, KY, USA

Aims and Scope

As the field of AI and Robotics continues to grow, the timely dissemination of emerging research and developments is increasingly important. This new book series serves as a publication venue for innovative technical contributions in AI and robotics, with coverage of both theoretical AI research and applied AI and robotics. The scope of the series includes, but is not limited to, titles in the areas of knowledge representation and reasoning, affective computing, deep learning/neural networks, natural language processing, AI safety, machine ethics, superintelligence, technological singularity, multi-agent systems, programming languages for AI, robot navigation, military, medical, humanoid robotics, autonomous robot, artificial consciousness, computer vision, pattern recognition, and other relevant topics which might be proposed by potential contributors.

Published Titles

The Virtual Mind: Designing the Logic to Approximate Human Thinking
Niklas Hageback

The Virtual Mind

Designing the Logic to Approximate Human Thinking

Niklas Hageback

CRC Press
Taylor & Francis Group
Boca Raton   London   New York

CRC Press is an imprint of the
Taylor & Francis Group, an **informa** business

A CHAPMAN & HALL BOOK

CRC Press
Taylor & Francis Group
6000 Broken Sound Parkway NW, Suite 300
Boca Raton, FL 33487-2742

© 2017 by Taylor & Francis Group, LLC
CRC Press is an imprint of Taylor & Francis Group, an Informa business

No claim to original U.S. Government works

Printed on acid-free paper

International Standard Book Number-13: 978-1-1380-5402-8 (Paperback)
                                        978-1-1380-5403-5 (Hardback)

Library of Congress Cataloging-in-Publication Data

Names: Hageback, Niklas, author.
Title: The virtual mind : designing the logic to approximate human thinking / Niklas Hageback.
Description: Boca Raton : Taylor & Francis, CRC Press, 2017. |
Series: Chapman & Hall/CRC artificial intelligence and robotics series | Includes bibliographical references and index.
Identifiers: LCCN 2017009374| ISBN 9781138054028 (pbk. : alk. paper) | ISBN 9781138054035 (hardback)
Subjects: LCSH: Artificial intelligence. | Logic. | Thought and thinking. | Other minds (Theory of knowledge)
Classification: LCC QA335 .H34 2017 | DDC 006.3--dc23
LC record available at https://lccn.loc.gov/2017009374

Visit the Taylor & Francis Web site at
http://www.taylorandfrancis.com

and the CRC Press Web site at
http://www.crcpress.com

Contents

Author

Niklas Hageback has extensive experience in the financial sector working at tier-one financial institutions and consulting firms, such as Deutsche Bank, KPMG and Goldman Sachs, where he held regional executive risk management and oversight roles in both Europe and Asia.

His current focus is on behavioural and cognitive psychology and is managing a portfolio of software start-up firms that are developing applications in those areas, in particular towards economics and finance. In 2014, he published the bestseller *The Mystery of Market Movements: An Archetypal Approach to Investment Forecasting and Modelling* (Bloomberg Press) in which he proposed an alternative methodology to forecasting trends and bubbles in financial markets. He is a frequent commentator on economic and political issues and operates a popular blog.

Introduction

I never came upon any of my discoveries through the process of rational thinking.

Albert Einstein (1879–1955)

The computer, sometimes considered an electronic counterpart to the mind, operates on Boolean logic; however, any approximation with human thinking is usually very poor. The structure of the mind seems to be working on a dialectic intrinsically more complex.

It seems odd that although we have now lived with machine-generated computation for more than 70 years, its underlying logic has still not advanced to more closely resemble human cognition. Why is it so?

How the human mind works has long been pondered upon; traces can be found among the earliest known writings, often in a religious context debating the soul, and leading up to the theories of the early twentieth century by Sigmund Freud and Carl Gustav Jung, among others, proposing and documenting an unconscious part of the mind.

The study of the mind, spanning from religion and philosophy to the various facets of psychology and in the current setting, neuroscience, have all concluded the existence of an unconscious part of the mind that functions distinctively differently from the conscious part. Thus, the holistic human mind forms an amalgamation of an unconscious and a conscious part, incorporating a structure for the decision-making process that operates in stark contrast to the standard logic framework.

Human thinking is a concoct of the largely rational reasoning by the conscious part of the mind of a skewed and truncated reality that is set by the reigning narrative to which a society conforms, and the particular logic of the unconscious that absorbs a broader sphere of reality. It interacts in accordance with a protocol, which often manifests in decision making that can be perceived as seemingly irrational but is far from it;

rather it follows a diverging schema that can be replicated and standardised into machine-generated thinking.

If attempting to replicate human thinking in the aspiration of developing a new computer architecture, the dual process theory provides a promising starting point, introducing two different kinds of thinking: one that is based on reasoning, often similar to Boolean logic, and a more intuitive form based on associations where previous perceptions serve as references. The dual process theory is an attractive proposition as we can in it recognise our own thought processes with the interspersed mix of rational thinking, and what often is labelled as 'gut feel' or intuition, where the path to our decision making cannot exclusively be described in terms of rationality but where emotions to an extent relaxes such properties. But where the dual process theory, in its various variations from William James up till more recently Daniel Kahneman's bestseller, *Thinking, Fast and Slow*, falls short is in amalgamating the timing in selection, and any dynamic constants, between these two systems of thinking. While in hindsight, the dual process theory can provide explanations to which of the two systems that dictated a particular outcome, there has as of yet been no comprehensive methodology presented on the prospects of forecasting the patterns of human thoughts.

But overlaying the dual process theory with Freudian and Jungian concepts, such as the abstract notion of the mind and the (collective) unconscious in a multidisciplinary approach, provides a promising framework for establishing a rule-based mechanism to replicate human thinking.

While the century-old theories of Freud and Jung might appear antiquated, recent findings in *neuropsychoanalytics* are breathing new life in them by confirming the existence of an unconscious playing an active part in decision making and how perceptions are interpreted. To Freud, the unconscious stored perceptions repressed by the conscious, and Jung theorised that the unconscious organised itself in innate patterns, *archetypes*, that activate when an accumulation of repressed perceptions overextends and starts to influence conscious thought patterns. How then to assess the contents of the unconscious? Empirical studies highlight that activities in the unconscious can be tracked through figurative language that indicates the themes around which the unconscious evolves. By extracting, categorising and statistically analysing these, it is possible to establish mechanistic rules and dynamic constants that are tested through a big data approach from public media, and with that standardisation and machine-generated thinking and thus the introduction of a new computer architecture becomes possible.

There is reality and then there is a social reality...

That humans truncate reality has long been an established fact. Every time époque operates by a different narrative and context that sets the overall cultural, political -isms, norms, mood entiments and moral ambience, such as the *Victorian* era.

Certain fixed ideas or themes will come to exist within this social reality, or *zeitgeist,* and carry strong emotional power to become the core of a belief system with rationality not applied on them but rather *within* them as they form the axioms which cannot be questioned. So, what in retrospect might appear to have been absurd statements or decisions made by someone were, in fact, highly rational, under the rigorous context and thought pattern dictated by the prevailing zeitgeist. A model of human thinking thus needs to be able to distinguish between reality and social reality.

The unconscious operates on a different logic...

In psychoanalysis, one of the most important characteristics of the unconscious is its aspiration towards symmetry, where concepts such as space, time and causality become one. It can be formulated through a suite of axioms that points out the anomalies from Boolean logic. These are the *principle of generalisation* and the *principle of symmetry*. Hence, the unconscious is marked by symmetry, where sameness is preferred, and its processes work with classificatory activity; they seek the similarities (associations) between objects, whereas for the conscious, distinguishing difference, or asymmetry, is the focus. These two logic systems work in a mixture, in what is termed *the bilogic thinking system*, which operates in a transcending manner where particular combinations of symmetrical logic and asymmetrical logic are morphed into bi-logic.

For whom is this book written?

This book, *The Virtual Mind: Designing the Logic to Approximate Human Thinking*, through an in-depth review in a multidisciplinary manner, takes aim at defining the makeup of the unconscious and conscious part of the mind and its integrated patterns.

From this perspective, the author proceeds with formulating a unique concept to designing the logic that approximates human thinking, which can be implemented on a variety of platforms to better forecast human behaviour.

Equipped with this insight, the reader will be facilitated with the knowledge to develop and produce a machine-generated virtual mind presented in a step-by-step blueprint manner. *The Virtual Mind* is therefore a must-read for anyone with an interest in the construction of the next generation of computer logic and artificial intelligence, and enhancing the understanding of the workings of the mind.

The book, *Virtual Mind: Designing the Logic to Approximate Human Thinking*, consists of six chapters:

- **Chapter 1: The Theory of the Mind.** This chapter provides the historical background to the various theories of the mind; from the ancient Greeks as well as considering the religious viewpoints on the characteristics of the human soul. In the early twentieth century, Sigmund Freud and Carl Gustav Jung made ground-breaking progress when they presented a structured approach to the human mind, including an unconscious part. Modern psychology has further elaborated on these perspectives, as have neuroscience, taking the anatomical view through the now ongoing large-scale projects to fully dissect the functioning of the brain. Other models of the mind, including the dual process theory, will also be dissected to draw out common generic features that are shared cross-disciplinary to give us what to date have been ascertained in our knowledge of the mind and its capabilities.

- **Chapter 2: There is Reality and Then There is (Social) Reality.** Why is it that humans blind out certain aspects of reality in a way that best can be described as an *Emperor's New Clothes* syndrome? There is an obvious discrepancy between *social reality* and *physical reality*; however, it seems what is being blinded out shifts over time and societies and cultures. To explore and further the discussion, the author will examine what is and what is not considered to be normal as it comes to serve as a distinguisher to what parts of reality are to be considered at any given time. As certain objects or topics become taboo and other become norms of society, and albeit hard to exactly pin down, together these define the social reality highlighting the preferences and providing the contemporary narrative. And while these concepts are well recognised by academia, they have fallen short in providing explanations on how social reality forms in structure, duration and content.

Psychological notions, such as groupthink, a tendency to conformity and Freudian defense mechanisms, all supports the creation of a social reality constrained by boundaries, which at times appears arbitrarily set. The social reality dictates to a large degree the perimeters that rational thinking is confined to operate within and any attempt to machine-wise replicate human thinking must be able to reflect that. This chapter presents a plausible hypothesis on how it emerges and it can be precisely defined, which will provide a foundation to distinguish social reality from the physical reality.

- **Chapter 3: How Does the Unconscious Part of the Mind Operate?** Following up on the previous chapter, perceptions dimmed by social reality, where do they go? We know through neuroscience that they are registered and stored in the unconscious, but although not recognised by the conscious part of the mind, can they still influence human decision making? If so, by which set of logic, if any, is this consciously unrecognised information arranged and operated by? Both Freud and Jung formulated rules and structures which the unconscious adhere to. Freud's proposition was later on picked up by the Chilean psychologist Ignacio Matte Blanco, who proposed a bi-logic schemata of the juxtaposition of conscious and unconscious that establishes the framework which the human mind functions by. This chapter aims to construe and present the algorithms which regulates the unconscious part of the mind.

- **Chapter 4: The Mind: The Meta-Model.** With the establishment of the restrictive arrangements that provide the social reality upon which the conscious part of the mind is applying its rationality and the specific logic of the unconscious part of the mind, this chapter outlines a meta-model over the interaction between these two parts that form the thought patterns of the human mind and serves as the basis for general decision making. The core modelling is to combine the opposing ideas from Freud and Jung to the dual process theory, thus offering an approach which brings the associative and the rule-based systems together into a holistic model with a protocol that governs the switches between conscious and unconscious logic depending on input and scenario. Provided with mechanistic rules and dynamic constants, tested through a big data approach from public media, they allow for standardisation and the introduction of a new computer architecture. With the meta-model available, the reader will better understand how trends and decisions seemingly, in fact, a strict set of rules, which can be simulated in a machine-generated fashion.

- **Chapter 5: An Object-Oriented Architecture Perspective in Designing a Virtual Mind.** There have over time been various attempts to design machines to reflect the way humans think, and *artificial intelligence*; incorporating techniques such as neural networks and fuzzy logic, is the main academic discipline and field of science under which such developments sort. The *Turing Test* has become one of the key features in determining how successful these efforts have been but is not the only one.

Various paths and theories have been explored; however, the attempts to simulate a virtual mind have been constrained by the architecture of the computer hardware; namely it operates exclusively on Boolean logic. And without relaxing this limitation and allowing for a completely new design, any attempt to replicate human thinking becomes less feasible. This chapter will, after a review of existing applications and design propositions evaluating their pros and cons, outline, based on the meta-model of the human mind in the previous chapter, an object-oriented architecture blueprint. This is presented in an easily available fashion, which permits the reader to attempt the implementation, development and elaboration of a computer logic better reflecting the mind than existing set-ups.

- **Chapter 6: Conclusions.** The concluding chapter summarises in a digested manner the meta-model for a virtual mind and draws attention to future areas of research.

This book forms part of a broader undertaking in designing and producing an electronic virtual mind; a new type of computer architecture that aims to augment, and to some extent replace, the need for a human mind in situations where traditional computer architecture often fails, for example, when trying to forecast (collective) human behaviour.

These efforts are supported by an online resource, *www.the-virtual-mind.com*, which provides frequent and regular updates on what the current themes are in terms of unconscious and conscious thought patterns and supportive analytics highlighting insights to emerging cultural, societal and economic trends that hold the propensity to break and reshape the reigning narrative.

CHAPTER 1

The Theory of the Mind

We must no more ask whether the soul and body are one than ask whether the wax and the figure impressed on it are one.

Aristotle (384 BC–322 BC)

THE EARLIEST THINKING ON THINKING

The function of the mind and especially how it relates to the body has been deliberated since the earliest days of man, and the debate is yet to be concluded. Strange as it might seem, something so essential in defining the human species still remains so elusive.

The concept of the mind has been studied from many perspectives: traditionally as part of religion and philosophy, in modern times having been directed towards psychology and more recently, neuroscience. What has and has not been part of the faculties that constitute the human mind has differed over time; however, a current definition of the mind reads:

> *the collective conscious and unconscious processes in a sentient organism that direct and influence mental and physical behavior*[1]

> *The elements of sentience being the brain, nerve processes, cognition and the motor and sensory processes.*

What is generally acknowledged is that the mind includes attributes such as perception, reason, imagination, awareness, memory, emotion and a faculty for exchanging information. However, the mind's full suite of properties is yet to be conclusively established.

Already during the time of the ancient Greeks, the distinction, *if any*, between the mind and the body was being contemplated and discussed. The Greek philosopher, Aristotle (384 BC–322 BC), regarded the soul, of which the mind was one constituent, as a part of the body, and when the body ceased to exist, so did the soul.[2] His Greek contemporary, Plato (427 BC–347 BC), however, suggested a divided mind and matter concept, where the soul (mind) survived the extinguishment of the body and therefore was immortal. He viewed the soul as universal and timeless, conforming to various recurring *forms*, or ideas, which represent the true reality unlike that which is experienced by the body.[3]

Over time, three schools of thought have come to the forefront: the *dualistic,* which considers the mind to exist independent of the brain; the *idealistic*, which deems that only mental phenomena exist; and the *materialistic*, in which the mind is the result of activities of the brain.[4]

The current focus of research, which is geared towards neuroscience and the man–machine relationship, has put the materialist perspective in the forefront. It has been evidenced that *certain* functions of the mind can be pinned to the brain, some of which we are consciously aware of, and others not. The study of patients with brain damages shows that injuries to specific parts of the brain result in impairments in functions seen as part of the mind. Experiments with drugs have also revealed brain–mind links; for example, sedatives reduce awareness, while stimulants do the opposite. But the advancements of neuroscience and genetics have not yet provided a comprehensive picture of how the brain produces and relates to various functions of the mind. For example, although some emotions can be directly related to certain brain structures, neuroscience still falls far from fully explaining emotions in terms of brain processes.[5]

An important concept in the debate is *qualia*: the individual subjective description of a perception or experience, such as describing an emotion, the level of pain, how something tastes, smells and so on. In essence, qualia means that it is difficult to objectively describe an experience, without it containing a subjective element that others can find difficult understanding and identifying with. Qualia brings the mind–body problem to its core, as it so far has not been possible to neuroscientifically explain the subjective 'making sense' of a certain experience and why it can differ from person to person.[6]

THE RELIGIOUS PERSPECTIVE

Most religions include the concept of *the soul*, sometimes synonymous with what we understand as the mind. The soul is typically considered as a non-physical substance with spiritual qualities existing separately from the body. The focus of most religions is to ensure that the soul survives the death of the body; hence being a distinct part, preferably remaining spiritually pure and enlightened, entering a Heaven rather than a Hell. And some religions, such as Buddhism, see the soul as reborn after death into another body, of whose qualities one's deeds in the previous life depend on.[7]

Up to the mid-1800s, the study of the soul from a Christian theological perspective, with its relationship to the human body and its fate in the afterlife, was the main concern in contemplating the structure of the mind. Concepts such as the *Holy Spirit* and *demons* came often to be assigned to human behaviour that appeared irrational. Mystical forces with clear religious connotations were considered as influences on the mind, and such forces were poetically described in the ancient Greek dramas and later on extensively worked into fictive literature such as that by William Shakespeare.

European philosophers of the seventeenth and eighteenth centuries studied the nature of these phenomena, and it was during this period that the term *unconscious* was coined. The view of the unconscious was as something externally induced, somehow forced against one's will and taking control of one's actions. And with the contemporary pondering upon that comes the questions that if humans are not always in control of their actions, the assumption of free will must be challenged and that opened up for a still ongoing debate not only from the philosophical perspective but also touching upon areas such as legal accountability.

SIGMUND FREUD

The Austrian psychologist, Sigmund Freud (1856–1939), took great interest in the function of the mind, especially the unconscious, came to revolutionise what up until then had been a quite narrow understanding of its make-up. Through his work with neurotic patients, he concluded that the formation of neurosis stemmed from the repression of emotions or desires, 'forbidden' by societal norms. According to Freud, these forbidden thoughts just did not go away but were stored in the unconscious and kept impacting behaviour. He worked with different techniques, eventually developing his own landmark *psychoanalysis* to draw out this repressed material from the unconscious into the conscious in order to help the patient free himself from neurosis. These insights drew Freud to the conclusion that the mind operates through psychic energy, or *libido,* and with the mind as a closed system, this energy level is constant, so by examining its flow it provides an understanding of any imbalances between the conscious and the unconscious. He claimed that the flow of the libido was dictated by two opposing forces: *the pleasure principle* and *the reality principle.* The *pleasure principle* triggers the impulse to seek immediate gratification of desires and urges. Freud saw the pleasure principle as the main drive of unconscious desires. The *reality principle* resides in the conscious part of the mind and reins in the impulses seeking immediate gratification. It represses socially unacceptable urges by redirecting the libido into more socially acceptable behaviour, such as artistic pursuits, through a process called *sublimation.* However, if the libido cannot be redirected towards acceptable alternative behaviours then neurosis begins to develop. Later on, Freud somewhat revised his theory and presented a third force, *the death drive*, which represents (self-)destructive tendencies that can spring to action if external circumstances create superfluous settings that the mind feels it does not correspond to its self-image, and thus needs to be rebalanced 'downwards'.[8,9,10]

In his essay from 1915, 'The Unconscious', Freud provided the following characteristics of the unconscious:

- It can be contradictory; opposing feelings or wishes can coexist. For example, you can feel love and hate for the same thing at the same time.

- Repressed thoughts or emotions are likely to return to the conscious in some form.

- The unconscious is timeless; its contents have no chronological order, the cause and effect relationship can be put out of play.

- In people with a mental disorder, it can thrust itself into the conscious and replace physical reality with psychic reality, such as fantasies, dreams and symbolism.[11]

With a perceived better understanding of the unconscious, Freud developed a holistic concept of the mind, in particular, to highlight the relationship between the conscious and the unconscious (see Figure 1.1). It is often depicted like an iceberg, with the conscious mind as the tip of the iceberg just above water level and the unconscious mind residing below the water level. It introduces the concept of a *preconscious* in between the conscious and the unconscious – a part of the mind whose contents are unconscious but not repressed, and can easily be recalled into the conscious. It is important to

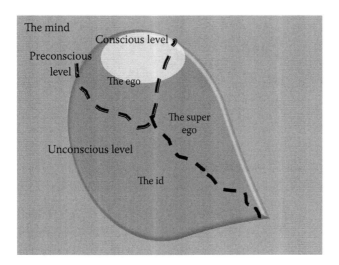

FIGURE 1.1   Freud's structural model of the mind.

note that Freud did not believe that, at the time, the conscious, preconscious and unconscious could be pinned down to physical locations in the brain, but rather were abstract phenomena – a mental model. The iceberg metaphor also highlighted his view that the preconscious and unconscious comprise the largest parts of the mind. Eventually, Freud reworked some earlier attempts of a model of the mind, and arrived at the *structural model*, containing the concepts of the *id*, the *ego* and the *superego*.

- *The id*, which is an innate function and resides in the unconscious, contains mankind's basic drives. It is a source of the libido, and acts along the lines of the pleasure principle.

- *The ego* resides in the conscious but partly also exists in the preconscious and unconscious. Influenced by the reality principle, it seeks to redirect drives from the id that are considered unacceptable.

- *The superego* is largely an unconscious function but with parts also in the conscious and preconscious. It serves as a kind of conscience, the reality principle, aspiring to comply with societal norms but also the ego's ambitions, and hence has a policing function over the drives of the id.[12]

CARL GUSTAV JUNG

Carl Gustav Jung (1875–1961) was a Swiss psychiatrist, and for a time one of Freud's closest disciples. At first, he subscribed to Freud's theories but over time started to diverge; among others, he saw Freud's view of the unconscious as incomplete. Whereas Freud viewed it as a dustbin for repressed emotions and desires rejected by the conscious, Jung saw it as comprising two distinct sections – a collective part and an individual part – and believed the unconscious played a more active role in the mind than what Freud proposed. He also dismissed Freud's models of the mind as arbitrary and too simplistic.[13]

Jung, who based his empirical work on patients and wider historical research, started to develop his own theory. While he shared Freud's view that the mind is divided into two parts, the conscious and the unconscious, which interact with each other, he saw additional dimensions of the mind: the personal and the collective.

Whereas the *personal conscious* is unique to each individual, the *collective conscious* is a form of public opinion, a distillation of the average person's cultural and moral values into a set of societal beliefs, norms, attitudes and mainstream political -isms. The values held in the collective conscious tend to be shared among the greater majority of individuals of the group and with the shared values generally superseding and repressing any conflicting personal beliefs or values.

The personal unconscious is defined by Jung as containing

> *lost memories, painful ideas that are repressed (i.e. forgotten on purpose), subliminal perceptions, by which are meant sense-perceptions that were not strong enough to reach consciousness, and finally, contents that are not yet ripe for consciousness.*[14]

Hence, the contents of the personal unconscious are unique to each individual much like that of the personal conscious.

Jung's definition of the personal unconscious comes close to Freud's view of the unconscious. However, Jung came to believe that something in Freud's model was missing and as his research into schizophrenia progressed, he started questioning the assumption that the human mind develops from a blank slate, *tabula rasa*, at birth. He hypothesised that innate, universal patterns existed in the unconscious – separate from the personal unconscious – and that these patterns could influence thought and behaviour. The concept of a collective unconscious was thus born.

Through studying his patients' and his own dreams and the review of diverse myths and sagas, these became the main conduits of exploring the unconscious. Jung identified recurring patterns of themes that seemed to exist collectively regardless of era, culture or geography – the *archetypes* (from the ancient Greek *arkhetupon*, which can be translated as an original model that can be copied). These archetypes would be representations of typical events in life, such as birth and death; characters, such as the father and the mother; and objects, such as the sun and fire, that any human collective will have to relate to and carries a symbolic as well as factual meaning.

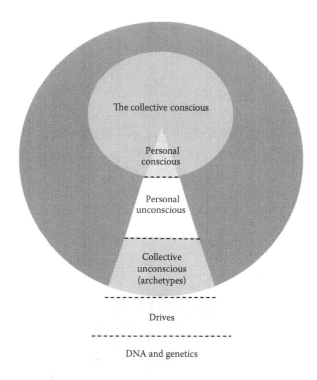

The collective conscious

Personal
conscious

- - - - - -

Personal
unconscious

- - - - - - - - - - -

Collective
unconscious
(archetypes)

- - - - - - - - - - - - - - - - - - -

Drives

- -

DNA and genetics

FIGURE 1.2   Jung's model of the mind.

He believed that archetypes are inherited with the brain structure and therefore biologically encoded (see Figure 1.2). Jung
saw them as being part of man since the earliest days, shaped by
evolution and representing the collective history of the human
species as the product of the constantly repeated experiences. It
was therefore apt that Jung referred to them as '*the two-million-
year-old man that is in us all*.'[15]

These archetypes generally lie dormant in the collective unconscious, in other words they are not actively influencing thoughts
and behaviours. But when a catalyst – an event or emotion in
conscious reality – creates a sufficiently strong psychological impetus, an archetype somehow related to that catalyst

awakes and begins to stimulate the conscious and alter perceptions. *Archetypal images* begin to appear, usually as symbols in visions, dreams, language and other forms of expressions, and eventually affect conscious thinking and prompt a mental context in which differing sets of actions are likely to be taken compared to the pre-archetype era in order to rectify the psychological unbalance and reduce the likelihood for neurosis. In sorts, these psychological forces work much like instincts, as certain situations make them trigger.

As archetypes reside in the unconscious, we are neither aware of their existence nor of their influence on our perceptions. As such they are an invisible force like that of a magnet attracting items in its direction. Even as they become activated, affecting behaviour and how we grasp reality, they remain obscure. But eventually they are detectable through observing the footprints they leave – the archetypal images. It is through the appearance of archetypal images in the conscious that we are able to infer the existence of the archetypes.

In Jung's own words

> *Archetypes are irrepresentable in themselves but their effects are discernible in archetypal images and motifs. Archetypes ... present themselves as ideas and images, like everything else that becomes a content of consciousness.*[16]

As Jung pointed out, concepts similar to archetypes appear in other academic disciplines and that list has since further expanded. And although there never have been attempts to formally connect them in a multidisciplinary approach, these forces do seem to share a number of characteristics:

- Innate in nature and part of the human psychological make-up

- Universal and recurring throughout humanity's different aspects over time
- Finite in number
- Generally differing from instincts in terms of being less constrained in their direct manifestations in response to catalysts[17]

THE DUAL PROCESS THEORY

Eventually, and after their demise, Freud's and Jung's theories fell out of fashion as they could not be backed by hard evidence and remained anecdotal and case study based in nature, and other theories of the mind emerged. What has since become the leading mind theory is the dual process theory, which assumes that human thinking consists of two distinct parts that interact in tandem and form a holistic structure.

These parts are an associative system that stems from the unconscious and a logic system that is conscious. From its first known reference by the American psychologist, William James (1842–1910), up till its most recent adaption in Daniel Kahneman's bestseller, *Thinking, Fast and Slow,* broadly, the common characteristics of the two systems can be outlined as in the table below:

Associative System	Logic System
Unconscious	Conscious
Implicit	Explicit
Automatic	Intentional
Joins perceptions through categorisations	Cause–effect consequential
Large capacity – independent of working memory	Small capacity – limited by working memory capacity
Fast	Slow
Brain's right hemisphere	Brain's left hemisphere

An important aspect to note is that the unconscious associative system is automatic i.e. one can neither 'turn it on' nor 'turn it off' unlike the conscious logic system that requires an explicit intentional act to commence.

But the dual process theory manages only to explain the decision-making path to a human action *a posteriori*, not *a priori*. No proponent of the dual process theory has yet managed to elaborate a methodology, which of the system that will be engaged in a specific decision-making scenario and if and when the associative system in some circumstances is likely to override the conscious logic thinking.

Nor has there been any explanations on whether there are any timing aspects in the selection, and any dynamic constants, between these two systems of thinking. And while the dual process theory has many merits in explaining human thinking, the lack of forecasting capabilities has reduced its capability in crafting practical applications.[18]

CURRENT THINKING

A century later, how does the theories of Freud and Jung stand up to current thinking and the findings of neuroscience? And, are there any evidence that supports the notion of two systems that forms holistic human thinking?

Neuroscience

As neuroscience now has taken the forefront in understanding the mind, projects to fully understand the functions of the brain have been launched. The *Brain Initiative* (Brain Research through Advancing Innovative Neurotechnologies) sponsored by the then President Barack Obama was announced in April 2013 with a view to acquiring a comprehensive understanding of the brain and its functions as well as gaining insights into brain disorders and psychological ailments such as depressions and schizophrenia.[19]

Also, the European Union have embarked on a similar endeavour, the *Human Brain Project*, aiming to develop an electronic prototype that replicates the human brain.[20]

However, one of the key issues that still persist and need resolve is the mind's subjective experience of the information and perceptions that it acquires and processes, such as the manner in which two persons could describe a colour or a particular taste distinctively different, which a computer operating on Boolean logic cannot replicate; the so-called 'hard problem' of consciousness, the aforementioned *qualia*.[21]

Evidence for the dual process theory have been found in numerous studies, where the different parts of the brain are activated depending on whether associative or analytical thinking are deployed. The right brain hemisphere processes data through unconscious, associative and automatic thought mechanisms, not in a fragmented manner but through configured patterns (or in Jungian terms something akin to archetypes). It also is not constrained by time or casual ordering of data. The left brain hemisphere handles the conscious, cause–effect analytical and intentional processing of data.[22,23]

There are efforts within neuroscience to try to combine the findings from the 'softer' psychoanalysis in order to better understand the extent of the mind and brain connect; this area of research has been labelled *neuropsychoanalysis*.

There are obviously elements of a cultural clash in attempting to marry these two perspectives; neuroscience with its strictly objective criteria *vis-à-vis* the Freudian approach of anecdotal evidence and essay-like descriptions.

But by trying to concoct these different views together, the mind–body construe is further elaborated and Freudian concepts could be pinned to specific structures or processes of the brain. Some of the current research areas within neuropsychoanalysis include the following:

- Defining the psychic energy, *the libido*, as a dopaminergic seeking system.

- Drives or instincts being mapped as emotions with the hypothesis that the brain has seven instinctual networks: *seeking*; *rage*; *fear*; *lust*; *care*; *panic/grief and play*, residing in pontine regions and projecting the cortex, with seeking being constantly active and the others striving for performance as required by the unconscious.[24]

View of the Unconscious

Psychology of today has to an extent moved away from Freud's view of the unconscious. The scientific community now believes the unconscious plays a more active part in certain decision-making processes, which is a marked shift from Freud's image of the unconscious as exclusively a dustbin of repressed material. In cognitive psychology, researchers are studying *implicit memory* – a type of memory in which previous experiences aid in the performing of a task without conscious awareness of these previous experiences. Psychologists have been investigating *priming* and *automaticity*, the ability to do something without being actively aware of it, such as repetitive menial tasks (e.g. riding a bicycle). Another area of research involves the *unconscious acquisition of information*. While a person is not consciously aware of himself absorbing information, these processes can still influence behaviour and decision-making processes. Empirical tests have shown that a person can act on information that was processed only in the unconscious *before* any awareness of that information or decision to act enters the conscious, for example, unconsciously determining patterns of sequences for various events. This appears to happen unintentionally, regardless of conscious goals and instructions. It seems to be a fully automatic unconscious process that is innate in all humans whatever their age, levels of intelligence, culture, education or other personal factors.[25,26]

If certain decision-making processes take place outside the conscious, and some seem even not consciously intended, how they are then accomplished remains an open question. Important to note is that more information and perceptions of reality are absorbed by the unconscious than the intake of the conscious part of the mind. The American cognitive psychologist George Miller (1920–2012) showed that the working memory can only hold seven items, give or take two, whereas the unconscious does not have such capacity constraints. As the conscious mind cannot absorb all perceptions, the unconscious stores this information where it can be retrieved later on. Hence, the information used for decision making by the conscious part is a truncated view of reality and because of that, psychologists are interested in how personal judgement and social behaviour operate outside awareness and conscious intent. While the unconscious acts as an independent agent outside the control of awareness, it also serves as a complementary partner to the conscious, manifesting itself usually in the form of intuition, gut feelings, eureka moments and other insights that we are not able to explain rationally.[27,28]

Another feature of the limited capacity of the conscious mind is that it must use schemas to work properly, something which the unconscious with its considerably larger capacity need not use to the same detailed extent, and thus can integrate information more efficiently.[29] Guy Claxton, an English cognitive scientist, has provided complimentary impetus to the dual process theory, through a component that forms the foundation of the unconscious thought theory, namely that conscious thought follows rule-based thinking, much in accordance with Boolean logic, and that the unconscious operates on associations, a symmetry of sorts either inherent or conditioned through cultivation by experience.[30] The view of the unconscious mind is not so much any longer whether it is involved in cognitive activities or not but rather how and how much.

Views of Archetypes

Jung's views of archetypes have gained some endorsement from more recent research; in *sociobiology*, it is assumed that our social conduct has developed through evolution and that it is inherited and affected by natural selection over the generations in the same way natural selection shapes physical features. That advantageous behaviour is eventually written into and embedded in the DNA. Sociobiologists and others now believe that humans (and animals) repeat acts that have proved beneficial from an evolutionary point of view. Therefore, it is possible in certain situations – those relating to evolutionary adaptation and survival – to forecast individual and collective human behaviour as they will fall back on these.[31]

Additional support for Jung's archetypal proposition comes from *epigenetics*; the study of how individual genes can be activated or deactivated through life experiences and/or the environment. It has been shown that the effects of behaviour or events in one lifetime can be genetically passed on to the next generation, causing a sudden 'evolutionary' change.

Rather than changing the DNA structure, a specific conduct or event switches on or off particular genes, and those genes remain switched on or off in the DNA passed on to the next generation with the capacity to affect both the offsprings' mental and physical health. The study of geographical regions with starvation or other severe environmental conditions has shown that these conditions can cause a change in genes that may be passed on within one generation, something quite contrary to previous perceptions that evolutionary changes take hundreds, or more, generations to take place.[32]

In his 2012 book, *The Neurobiology of the Gods*, Erik D. Goodwyn examines the most up-to-date evolutionary and cognitive neuroscientific research, seeking a way to understand archetypes in terms of brain physiology.[33] He documents extensive empirical studies that point to a neuroscientific basis for archetypal patterns. However, others, such as Christian Roesler in an article in the *Journal of Analytical Psychology*, 'Are Archetypes Transmitted More by Culture Than Biology?' argue that epigenetics in fact raises questions about many of Jung's basic assumptions.[34] New findings in neuroscience provide more and more insight into the functions of the brain and links to the hitherto abstract notion of the mind. While this area of study is still considered a frontier science, regardless of how one choses to label them, there seems to exist a broad scientific agreement that we have a kind of genetic memory, a tendency hard-coded into our DNA to respond in certain ways to certain stimuli. The notion that at birth our mind is a blank slate, *tabula rasa*, has been discarded and proven erroneous.

CONCLUSIONS

The millennia old probe of gauging the functionality of the mind has recently taken a new tack; with the advancement of neuroscience a better understanding of the brain and its functions is emerging. Progress has been rapid and it is expected to continue through the launched large-scale projects with a view to gaining a comprehensive understanding of the brain–mind relationship and any attempts to electronically replicate it.

What the shared features are between the Freudian and Jungian models and the dual process theory are the existence of an unconscious part that influences human actions and serves as a repository of information gathering. The unconscious operates on associations that are held together through categories, which in Jungian parlance is referred to as archetypes.

Hence, we know that an electronic counterpart to human thinking cannot operate on Boolean logic alone as there are subjective elements in the perceptions and constraints in the view of reality that has so far not been possible to artificially reproduce. Why is it that we chose to ignore sometimes obvious parts of reality in what can only be labelled *The Emperor's New Clothes* syndrome? The complexity of the mind is further added by the interaction between the conscious and unconscious parts of the mind and that operating protocol remains to be fully described and formalised.

The following chapters will delve into the mechanisms that adjust for the conscious part of the mind's skewed perception of reality, which leads to a subjective perspective in contrast to an expected rationality, especially when studied in hindsight.

The unconscious and the rules it functions by will also be discussed and its impact on human behaviour. These chapters will be consolidated into a holistic model and detailed through a schema that proposes a blueprint of an electronic replication of the human mind's structure and functionality.

ENDNOTES

1. *The American Heritage Dictionary of the English Language.* http://ahdictionary.com/ (accessed 1 January 2017).
2. The Internet Classics Archive. http://classics.mit.edu/Aristotle/soul.html (accessed 1 January 2017).
3. *Online Etymology Dictionary.* 2001. www.etymonline.com (accessed 1 January 2017).
4. Kim, J. 1995. Problems in the philosophy of mind. In *Oxford Companion to Philosophy*, edited by T. Honderich. Oxford: Oxford University Press.
5. Computational Neuroscience Research Group. Waterloo Centre for Theoretical Neuroscience. http://compneuro.uwaterloo.ca/index.html (accessed 1 January 2017).
6. Kriegel, U. 2014. *Current Controversies in Philosophy of Mind.* New York, NY: Taylor & Francis, p. 201.
7. *Encyclopædia Britannica.* Soul. http://www.britannica.com/ (accessed 1 January 2017).
8. Sheehy, N; Forsythe, A. 2013. Sigmund Freud. In *Fifty Key Thinkers in Psychology.* London: Routledge.
9. Freud, S. 1960. *Group Psychology and the Analysis of the Ego.* New York: Bantam Books.
10. Freud, S. 1987. Beyond the pleasure principle. In *On Metapsychology.* Middlesex: Penguin.
11. Freud, S. 2005. *The Unconscious.* London: Penguin.
12. Freud, S. 1990. *The Ego and the Id.* New York: W. W. Norton & Company.
13. Dunne, C. 2002. *Carl Jung: Wounded Healer of the Soul: An Illustrated Biography.* London: Continuum International Publishing Group.
14. Jung, CG. 1981. The archetypes and the collective unconscious. In *Part 1: The Collected Works of C. G. Jung*, translated by RFC Hull. Princeton, NJ: Princeton University Press, Vol. 9, 2nd edition, par. 103.
15. Stevens, A; Rosen, DH. 2005. *The Two Million-Year-Old Self*, Carolyn and Ernest Fay Series in Analytical Psychology. College Station, TX: Texas A&M University Press.

16. Jung, CG. Concerning the archetypes and the anima concept. In *Part 1: The Collected Works of C. G. Jung*, translated by RFC Hull. Princeton, NJ: Princeton University Press, Vol. 9, 2nd edition, par. 136.

17. Ibid.

18. Kahneman, D. 2011. *Thinking, Fast and Slow*. New York: Farrar, Straus and Giroux, 1st edition.

19. The Brain Initiative. https://braininitiative.nih.gov/index.htm (accessed 1 January 2017).

20. Human Brain Project. https://www.humanbrainproject.eu/ (accessed 1 January 2017).

21. Graziano, MSA. 2013. *Consciousness and the Social Brain*. New York: Oxford University Press.

22. Sternberg, RJ; Leighton, JP. 2004. *The Nature of Reasoning*. Cambridge University Press, Cambridge, MA, p. 300.

23. De Neys, W. 2006. Dual processing in reasoning: Two systems but one reasoner. *Psychological Science* 17 (5): 428–433.

24. Schwartz, C. June 28, 2015. Tell it about your mother – Can brain-scanning help save Freudian psychoanalysis? *New York Times*. http://www.nytimes.com/2015/06/28/magazine/tell-it-about-your-mother.html (accessed 1 January 2017).

25. Dijksterhuis, A; Smith, PK; van Baaren, RB; Wigboldus, DHJ; Bargh, JA; Morsella, E. 2008. The unconscious mind. *Perspectives on Psychological Science* 3 (1): 73–79.

26. Stein, DJ. 1997. *Cognitive Science and the Unconscious*. Arlington, VA: American Psychiatric Publishing.

27. Carlin, F. 2007. Gut Almighty. *Psychology Today* 40 (3): 68–75.

28. Dijksterhuis, A. 2007. Think different: The merits of unconscious thought in preference development and decision making. *Journal of Personality and Social Psychology* 87 (5): 586–598. http://domed.de/domed_downloads/Dijksterhuis%20-%20The%20Merits%20of%20Unconscious%20Thought%20in%20Preference%20Detection%20and%20Decision%20Making.pdf (retrieved 1 January 2017).

29. Dijksterhuis, A; Nordgren, LF. 2006. A theory of unconscious thought. *Perspective on Psychological Science* 1 (2): 95–109. http://

www.alice.id.tue.nl/references/dijksterhuis-nordgren-2006.pdf (retrieved 1 January 2017).

30. Claxton, G. 2006. *The Wayward Mind: An Intimate History of the Unconscious.* London, United Kingdom: Little Brown Book Group.
31. Wilson, EO. *Sociobiology: The New Synthesis.* Cambridge, MA: Belknap Press of Harvard University Press, 25th Anniversary Edition.
32. Francis, RC. *Epigenetics: How Environment Shapes Our Genes.* New York: W. W. Norton & Company.
33. Goodwyn, ED. *The Neurobiology of the Gods: How Brain Physiology Shapes the Recurrent Imagery of Myth and Dreams.* New York: Routledge.
34. Roesler, C. 2012. Are archetypes transmitted more by culture than biology? *Journal of Analytical Psychology* 57 (2): 223–246.

BIBLIOGRAPHY

Carlin, F. 2007. Gut Almighty. *Psychology Today* 40 (3): 68–75.

Claxton, G. 2006. *The Wayward Mind: An Intimate History of the Unconscious.* London, United Kingdom: Little Brown Book Group.

Computational Neuroscience Research Group. Waterloo Centre for Theoretical Neuroscience. http://compneuro.uwaterloo.ca/index.html (accessed 1 January 2017).

De Neys, W. 2006. Dual processing in reasoning: Two systems but one reasoner. *Psychological Science* 17 (5): 428–433.

Dijksterhuis, A. 2004. Think different: The merits of unconscious thought in preference development and decision making. *Journal of Personality and Social Psychology* 87 (5). http://domed.de/domed_downloads/Dijksterhuis%20-%20 The%20Merits%20of%20Unconscious%20Thought%20in%20 Preference%20Detection%20and%20Decision%20Making.pdf (retrieved 1 January 2017).

Dijksterhuis, A; Nordgren, LF. 2006. A theory of unconscious thought. *Perspective on Psychological Science* 1 (2). http://journals.sage-pub.com/doi/abs/10.1111/j.1745-6916.2006.00007.x?journalCo de=ppsa (retrieved 1 January 2017).

Dijksterhuis, A; Smith, PK; van Baaren, RB; Wigboldus, DHJ; Bargh, JA; Morsella, E. 2008. The unconscious mind. *Perspectives on Psychological Science* 3 (1): 73–79.

Dunne, C. 2002. *Carl Jung: Wounded Healer of the Soul: An Illustrated Biography*. London: Continuum International Publishing Group.

Encyclopedia Britannica. Soul. http://www.britannica.com/ (accessed 1 January 2017).

Francis, RC. 2012. *Epigenetics: How Environment Shapes Our Genes*. New York: W. W. Norton & Company.

Freud, S. 1959. *Group Psychology and the Analysis of the Ego*. New York: Bantam Books.

Freud, S. 1987. Beyond the pleasure principle. In *On Metapsychology*. Harmondsworth: Penguin.

Freud, S. 1990. *The Ego and the Id*. New York: W. W. Norton & Company.

Freud, S. 2005. *The Unconscious*. London: Penguin.

Goodwyn, ED. 2012. *The Neurobiology of the Gods: How Brain Physiology Shapes the Recurrent Imagery of Myth and Dreams*. New York: Routledge.

Graziano, MSA. 2013. *Consciousness and the Social Brain*. New York: Oxford University Press.

Human Brain Project. https://www.humanbrainproject.eu/ (accessed 1 January 2017).

Jung, CG. 1981a. The archetypes and the collective unconscious. In *Part 1: The Collected Works of C. G. Jung*, translated by RFC Hull. Princeton, NJ: Princeton University Press, Vol. 9, 2nd edition.

Jung, CG. 1981b. Concerning the archetypes and the anima concept. *Part 1: The Collected Works of C. G. Jung*, translated by RFC Hull. Princeton, NJ: Princeton University Press, Vol. 9, 2nd edition.

Kahneman, D. 2011. *Thinking, Fast and Slow*. New York: Farrar, Straus and Giroux, 1st edition.

Kim, J. 1995. Problems in the philosophy of mind. In *Oxford Companion to Philosophy*, edited by T. Honderich. Oxford: Oxford University Press.

Kriegel, U. 2014. *Current Controversies in Philosophy of Mind*. New York, NY: Taylor & Francis, p. 201.

Online Etymology Dictionary. 2001. www.etymonline.com (accessed 1 January 2017).

Roesler, C. 2012. Are archetypes transmitted more by culture than biology? *Journal of Analytical Psychology* 57 (2): 223–246.

Schwartz, C. June 28, 2015. Tell it about your mother – Can brain-scanning help save Freudian psychoanalysis? *New York Times.* http://www.nytimes.com/2015/06/28/magazine/tell-it-about-your-mother.html (accessed 1 January 2017).

Sheehy, N; Forsythe, A. 2013. Sigmund Freud. In *Fifty Key Thinkers in Psychology.* London: Routledge.

Stein, DJ. 1997. *Cognitive Science and the Unconscious.* Arlington, VA: American Psychiatric Publishing.

Sternberg, RJ; Leighton, JP. 2004. *The Nature of Reasoning.* Cambridge, MA: Cambridge University Press, p. 300.

Stevens, A; David, HR. 2005. *The Two Million-Year-Old Self (Carolyn and Ernest Fay Series in Analytical Psychology).* College Station, TX: Texas A&M University Press.

The American Heritage Dictionary of the English Language. http://ahdictionary.com/ (accessed 1 January 2017).

The Brain Initiative. https://braininitiative.nih.gov/index.htm (accessed 1 January 2017).

The Internet Classics Archive. http://classics.mit.edu/Aristotle/soul.html (accessed 1 January 2017).

Wilson, EO. 2000. *Sociobiology: The New Synthesis.* Cambridge, MA: Belknap Press of Harvard University Press, 25th Anniversary edition.

CHAPTER 2

There Is Reality and Then There Is (Social) Reality

Men han har jo ikke noget paa!
(But he isn't wearing anything at all!)

Keiserens Nye Klær (The Emperor's New Clothes)
by Hans Christian Andersen (1805–1875)

WHAT PREVENTS HUMANS FROM PERCEIVING REALITY AS IT REALLY IS?

Most theories in social science operate on the assumption of *the rational man hypothesis*; the information we acquire and absorb are processed in accordance with a logic system, typically Boolean logic. Any output will therefore, in its simplest form, be governed and calculated by two main laws: *the law of contradiction*, which states that it is impossible that *p* and *-p* can exist at the same time, hence one side of a contradiction must be false; and *the law of the excluded third*, according to which there cannot be anything between to be and not to be.[1]

But clearly, this is not how the human mind works, as in such case a computer could be programmed with the mechanistic rules that constitutes Boolean logic, and with a high degree of accuracy could forecast human behaviour in any given situation, if provided with the same input and context. There is also a sliding scale of preferences and inclinations each individual subscribes to, which further complicates how a set of mechanic decision-making rules would be set up i.e. the preference of a certain product over another but only in a specific situation and perhaps only to certain amounts, in an alternate setting, the choice would be different, even radically different. Obviously, that would mean a lot of rules and conditions but most scenarios could be covered and developed into programming code, and the output would be machine-generated thinking coming close to that of human thinking.

However, as the academic disciplines psychology and biology differ from social sciences with its rational man hypothesis, in that they have long recognised that humans curtail their experience of reality, there are boundaries on how perceptions are filtered and processed, and concepts such as normal, norms, taboo and zeitgeist help control whether the proverbial cup should at times be seen as half full and at other times as half empty.

If the way the world viewed is conditional to subjective constraints, it amply influences how the human mind interprets information. And without understanding the contents of these constraints and their reach, any attempt to project future human actions in most situations based on Boolean logic, or strict rationality, becomes futile. Whereas this subjectivity will hold distinct unique individual traits, an attempt to replicate it should start with the more common traits, with that sets the boundaries for what is considered normal in society at large. What is more is that these boundaries change over time. The challenge therefore is to define the range that provides the zeitgeist, the formative acceptable considerations of reality, regardless of being individually accepted or not.

NORMAL AND NORMS

What is normal is a truly ambiguous notion; in a cannibalistic society eating human flesh is perfectly normal. As it is so dynamic, it will vary depending on setting and as difficult such a fluid concept is to pin down, it is still pivotal to understand as individuals and the collective behaviour will be regulated by it, whether codified into formal legislation or not, because it becomes such a major influence on how we conduct ourselves. What is normal often only becomes obvious when something *abnormal* occurs. Abnormality is often linked to psychological ailments and the descriptions thereof are sometimes used as attempts to define what is outside the boundaries of normality, broadly assuming everything else as normal. For instance, homosexuality was for long considered abnormal in most Western societies, but has over the last few decades been redefined as an acceptable practice and thus falling within the boundaries of normal.[2]

The confines of normal is often, but not always, defined through *norms*; these include values, customs and traditions to which a collective subscribes. Norms can be arranged as laws but more often they develop informally as routines to control behaviour for some reason deemed harmful.[3]

In academia, there have been attempts to formulate the concept of normal; among others, the French sociologist Émile Durkheim (1858–1917) proposed that recurring behaviours among the majority of the population, which he labelled *social facts,* would represent normal and to the point where activities not considered acceptable however frequent, such as crime, could also be part of what is normal. Even if the criminal behaviour of an individual might be perceived as abnormal or unacceptable on the collective level, it averages out and becomes part of an everyday occurrence. Hence, when attempting to forecast human behaviour, a distinction needs to be made between the individual and the collective levels. Such divergence between the individual's normal and the society's normal can manifest in *pluralistic ignorance*; people pay lip service to the norms that society want their citizens to adhere to but that they privately abhor or vice versa. If the contradiction between the privately held norms and the societal becomes too wide, too harshly enforced or too prolonged, at the expense of the ones privately held, the risk for cognitive dissonance will increase; this in itself can lead to psychological disturbances in the form of neurosis.[4]

One approach to define normal is through assessing the statistical frequency and study trends over time to determine when the contents of normal changes and what they are. The normal and norms, therefore, produce a reality different from the actual reality, *the social reality.* It is this social reality that forms the basis for the rational decision making. Society will filter out what does not fall within normal and make projections and apply Boolean logic on the social reality rather than the actual reality. This has implications, among others, on the language that is set to describe reality and often becomes evident in the constant rewriting of historical events in each new time epoch.[5]

The study on how norms are established is a well-researched area, ranging from Durkheim proposing that norms are the statistically normal behaviour, whether morally agreeable or not, to more recently Cristina Bicchieri's work on strategies on how to introduce norms to elevate righteous behaviour. But the research remains silent on one key point; why does norms and their zeitgeist then not stay the same? Psychoanalysis do provide an answer, over time, norms have a tendency to become fixated into rigid formality in which the original purpose of the norm and psychological balance they were set to ensure no longer can be met. These unbalances trigger the neurotic behaviour and other psychological imbalances that Freud and Jung described and studied.[6]

ZEITGEIST

The collation of norms, consolidated together with statistically frequent items and topics for the specific time era considered as normal, brings with it the creation of a general societal mood referred to as *zeitgeist*; a German word that broadly translates as '*the spirit of the age*'.

Zeitgeist is, as its underlying determinants, a somewhat elusive concept, that provides a narrative and context to the overall cultural, political -isms, intellectual, mood sentiment and/or moral ambience for the particular time époque and collective group, often a society or culturally closely related countries. Certain fixed ideas or themes will come to exist within the zeitgeist, these carry strong emotional power and they become the core of a belief system, and rationality is not applied on them but rather within them as they form the axioms that cannot be questioned. These fixations become an embedded part of the collective's perception of reality. They start to trigger actions and behaviours to synchronise and align with the gist of the zeitgeist.

So, what in retrospect might appear to have been absurd statements or decisions made by someone while seized by the zeitgeist were, in fact, highly rational, under the rigorous context and thought pattern dictated by it.

Zeitgeists seem to govern which reality to focus on and what to consciously disregard but unconsciously record; in other words, turning a blind eye to features of reality or considering reality from a heavily skewed perception.

This is the power of the zeitgeist, the discrepancies between psychological illusions that perceive reality and reality itself, leading to a refusal to observe what is in front of us. Any critique against the validity of these axioms will be met by much fierce and emotional resistance, like that of questioning the beliefs of a religious person from a rational perspective. A weakening of the particular zeitgeist can be gauged through a lessened emotional response against the critique of these axioms.[7,8]

If elements of reality do not corroborate with the world view provided by the zeitgeist, these are either ignored or distorted and interpreted not to conflict with the zeitgeist. However, if these elements become too overwhelming and seem to threaten the foundation of societal values, they are not only frowned upon by a conformist public but even be declared 'illegal knowledge', and as a case in point, out of many, was the Galileo affair and the Vatican on the view of earth's role in the planetary system.

But over time, norms have a tendency to become fixated into rigid formality in which the original purpose of the norm and psychological balance they were set to ensure can no longer be met so that more and more aspects of reality have to be filtered out to not confront the conformity of the zeitgeist.

Reality eventually evolves, and activities once deemed harmful and needed to be controlled through norms can no longer be considered so, and keeping the norms in place risks creating unhealthy discrepancies between human instincts and the zeitgeist.

But the notion of instincts has since long been disregarded in contemporary political science and economics. Humans are considered to be governed by rationality alone, and thus political and economic systems are designed to be optimised for rational decisions. A recalibration of the zeitgeist is often the cause of turbulent times, even violent conflicts as the rational systems rarely are designed to facilitate the need of such changes.

That psychological imbalances lead to repression of instincts have been thoroughly researched on the individual level and its symptoms are well recognised, namely neurotic behaviour and conceding evidence support that what applies on the individual level is also applicable for the greater collective but with the difference that the societal neurosis and aggressions operate on a vastly larger magnitude.[9]

TABOO

Taboos are defined as items relating to human activities that are prohibited or strongly condemned, based on reigning societal or moral and religious convictions. As a rule of thumb, taboos are identified through the notably strong reaction to a particular word or theme triggering a typically heated emotion. In the

Victorian era, the noted taboo was anything related to sexuality. Another example is the swift change of discussion topic or other typical displacement activities that signal a disturbance in response to the taboo word or the use of euphemisms to neutralise the contentiousness. Taboos also cover the public discourse, what can and cannot be debated and is often seen as items considered inappropriate by political correctness. As with what is considered normal, what is seen as taboo changes over time and will differ depending on cultural contexts. There seems to be very few items that are perpetually seen as taboo or at least the scope of a certain taboo can vary widely.[10]

What is then the difference between taboos and norms? One interpretation is that taboos are strong norms; norms that are sufficiently strong that they may be viewed as sacred and breaching them will be castigated through severe social sanctions.

But the implications of taboo are wider than just being considered a more extreme form of norms; taboos are sometimes referred to as doing the *'unthinkable'*.

Even thinking about violating a taboo becomes problematic. Under this interpretation, a taboo is a form of a censor on thinking or in Freudian parlance; *the superego* that governs not just human behaviour but also its thoughts and in that sense taboos exists to prevent unconscious impulses from manifesting.[11]

CONFORMITY AND GROUPTHINK

What then are the psychological mechanisms that provide the 'glue' to attach the general public to the norms of a zeitgeist? Conformity is the act of aligning one's attitudes, norms and behaviour to that of the group's.[12] The reason for conformity is often based on a desire to seek security within the group rather than risk being a social outcast. These norms therefore form within groups that share some sort of commonality, whether of nationality, ethnicity, religion, etc.

By seeking conformity, individuals can come to revert to self-deception and force themselves to consent to the prescribed social reality as presented to them through the zeitgeist, in effect, this peer pressure causes *groupthink*.[13]

This aspiration for harmony tends to come at a cost; as the curtailed view of reality leads to inadequate decision making although at the time it is seen as entirely rational at least to the extent to what everyone outspokenly endorses.

The stronger the propensity for groupthink is, the harsher tends the deviators from the conformist views to be treated.

To assess whether groupthink is in place, the American psychologist Irving Janis (1918–1990) crafted a suite of criteria that would indicate the existence of groupthink in the collective, whether as society, culture, religion or others.

Type I: Overestimations of the group – its power and morality

- Illusions of invulnerability creating excessive optimism and encouraging risk taking
- Unquestioned belief in the morality of the group, causing members to ignore the consequences of their actions

Type II: Closed-mindedness

- Rationalising warnings that might challenge the group's assumptions
- Stereotyping those who are opposed to the group as weak, evil, biased, spiteful, impotent or stupid

Type III: Pressures towards uniformity

- Self-censorship of ideas that deviate from the apparent group consensus
- Illusions of unanimity among group members; silence is viewed as agreement
- Direct pressure to conform placed on any member who questions the group, couched in terms of 'disloyalty'
- Mind guards – self-appointed members who shield the group from dissenting information

Janis also highlighted three antecedent conditions to groupthink:

- High group cohesiveness
  - Deindividuation: group cohesiveness becomes more important than individual freedom of expression
- Structural faults
  - Insulation of the group
  - Lack of impartial leadership
  - Lack of norms requiring methodological procedures
  - Homogeneity of members' social backgrounds and ideology
- Situational context
  - Highly stressful external threats
  - Recent failures
  - Excessive difficulties on the decision making task
  - Moral dilemmas[14]

The level of adherence to these key elements decides the degree to which contrasting or directly conflicting views outside the realms of conformity are allowed by the governing discourse.

DEFENSE MECHANISMS

Sigmund Freud, and even more so his daughter Anna Freud (1895–1982) in her own right a renowned psychoanalyst, wrote extensively about the processes which excluded the parts of reality not deemed as being part of the zeitgeist; they referred to these as *defense mechanisms*.

Their typical function is to protect the conscious part of the mind from unwelcomed perceptions that will not corroborate within the acceptable, as such they are directed into the unconscious with a view to prevent mental discomfort or angst.

However, if it is allowed to linger on in ever-increasing numbers, it can be the cause of neurosis.

The capacity of the unconscious part of the mind to absorb a considerably larger amount of perceptions than that what is being consciously digested, has been verified through a number of empirical studies.[15] The Freuds identified a number of defense mechanism with some of the most important of them being the following:

- *Repression*: An impulse that forces something emotionally painful to be forgotten and consigned to the unconscious. Because the psychological issue has not been dealt with or resolved by the conscious, it continues to affect the person unconsciously and can eventually develop into mental disorders. For example, a person who represses memories of abuse experienced during childhood is likely to have difficulties forming relationships until the repressed feelings have been addressed.

- *Suppression*: A conscious act to forget something in an attempt to cope with a troubling situation. As suppression only pushes stressful thoughts or emotions into the preconscious, it is relatively easy to retrieve them later and work on accepting or resolving them. Suppression is similar to repression but the unpleasant feelings are pushed down consciously rather than unconsciously, and to the preconscious rather than the unconscious, and are therefore easier to bring back and address at a later stage.

- *Projection*: A person or group unconsciously projects thoughts or feelings not tolerated by the zeitgeist onto others, thereby creating scapegoats for particular issues.

- *Condensation*: Several concepts, including a taboo theme are blended with other non-threatening concepts and repressed into the unconscious. This produces a single symbol to represent the combined components. This symbol becomes a figurative form to represent the subdued impulses.

- *Denial*: A person or group denies reality by pretending it does not exist, like the deluded ruler and his subjects in the tale of the 'Emperor's New Clothes'.

- *Displacement*: The redirecting of feelings or actions from a dangerous outlet to a safer one; for example, a person yelling at his secretary after being reprimanded by his boss.

- *Rationalisation*: To seek a rational explanation or justification for upsetting actions or behaviour caused by factors too unpleasant to acknowledge; for example, a student who fails a test and rationalises that it was the result of poor teaching and not his lack of preparation. Rationalisation protects the ego by avoiding or circumventing the true reason for events or behaviour, whether or not controllable.

- *Reaction formation*: The disguising of beliefs or impulses considered unacceptable by the exaggerated expression of opposite beliefs or impulses. For example, some men unhappily possess homosexual tendencies and suppress these tendencies; they then project instead a hatred of homosexuals; they attack in others what they hate in themselves.

- *Regression*: The return to a previous stage of mental development in a situation of adversity, such as an adult acting childishly in response to a problem.

- *Sublimation*: The acting out of unacceptable impulses in a socially acceptable way, such as by finding outlets for the libido in cultural or intellectual pursuits.[16]

The existence of something akin to the defense mechanisms that Freud proposed, in particular suppression and repression, where perceptions not in congruence with acceptable norms are coerced into the unconscious have been acknowledged through more recent studies. Research points to a link between psychical ailments and a repressive personality, the ones with a tendency to avoid expressing emotions either intentionally or unconsciously when dealing with situations of distress by suppressing/repressing them.[17] Manifestations like neurosis and depression are the bodily symptoms from the unconscious that the conscious part of the mind cannot verbally formalise.[18]

THE SOCIOLOGICAL VIEWS

The fact that humans are not acting rationally has been accounted for in some sociological perspectives. In academia, the *rational man theory* has come to serve as a standard assumption, and it is an elegant, albeit highly unrealistic, assumption allowing to formulate mathematically exact outcomes in determining economic or social preferences.

As an alternative, however less adopted and evolved for modelling purposes, is the concept of *bounded rationality*, formulated by the American economist and sociologist Herbert A. Simon (1916–2001) and later on elaborated by the Israeli economist Daniel Kahneman.

In effect, it states that man is only partially rational in his decisions and is irrational in other parts of the decision making; examples are very complex decisions when one would typically fall back to rules of thumb or other short-cut approximations and thereby greatly simplify reality. However, as its proponents' difficulties in formulating generic rules for when bounded rationality would apply and when not, it has mainly been applied ex-post rather than ex-ante in explaining why a decision did not follow the expected path of rationality.[19,20]

Another concept eluding to similar functionalities as zeitgeist is the narrative in the public debate, or *discourse*, governing what can and cannot be debated and often also restricting who can participate. An entire strain of philosophy has been devoted to discourse analysis, and one of the key names was that of the French philosopher Michel Foucault (1926–1984). Foucault pointed out that the theme of the discourse dictates what is to be considered as contemporary dogma and that forms the slogans, even platitudes, to which individuals and society at large need to adhere to, and if challenged by anyone comes with the risk of becoming a social outcast. This discourse analysis not only takes aim at the themes and its meanings but also on the power relationships it highlights and aspires to maintain. Language is, therefore, seen as a tool for power and what thematic language is being used supports a certain ideology upon which the collective is expected to subscribe to. Much of the focus is on how authorities are broadcasting messages towards their underlings, exhibiting their dominance and demanding obedience.[21]

A Foucauldian discourse analysis broadly consists of four steps with the view to identifying the rules on:

1. How those statements are created

2. What can be said (written) and what cannot

3. How spaces in which new statements can be made are created

4. Making practices material and discursive at the same time[22]

How these discourses appear in the first place and the content they incorporate are never firmly established by Foucault et al. It is simply referred to as a controlling instrument for power that either seems to bestow authorities almost exogenously and then being wilfully constructed. This vagueness in explaining source and origin becomes obvious as the Foucauldian discourse analysis struggles with explaining shifts in discourse and contests of power between competing groups.

A somewhat similar theory has been presented in behavioural economics, labelled as *framing*, by the Israeli economists Amos Tversky (1937–1996) and Daniel Kahneman. Through experiments they showed that the way a problem is being framed and presented can affect what decision is made, and it can distort the assumed rationality and be used to game and manipulate behaviour and outcomes in a desired way. However, also with the concept of framing its proponents have failed to convincingly explain why it works sometimes and other times not.[23]

THE SHIFTING SOCIAL REALITY

The shift over time of social reality can be described through the sequence presented in Figures 2.1 to 2.3.

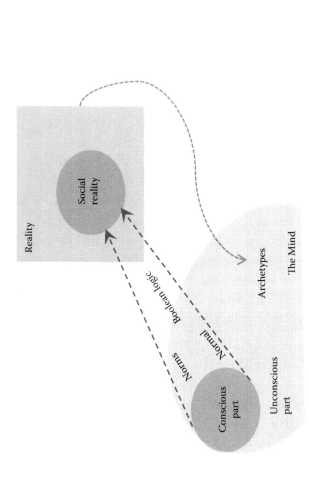

FIGURE 2.1 The conscious part of the mind's view of reality is constrained through a suite of norms that together with 'normal' behaviour forms a social reality of what is being included and excluded from the actual reality. Facts not concurring with the zeitgeist are disregarded, however still absorbed by the mind and stored in the unconscious part of the mind.

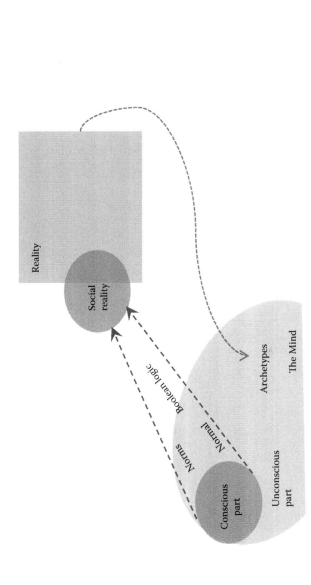

FIGURE 2.2   As reality changes over time, due to a paradigm shift or other evolvements, a rigid set of norms prohibits changes to social reality and the zeitgeist it represents, with most part of reality being ignored or viewed as taboo. More and more of the decision making within the context of the zeitgeist becomes too detached from reality and simply too abstract in nature.

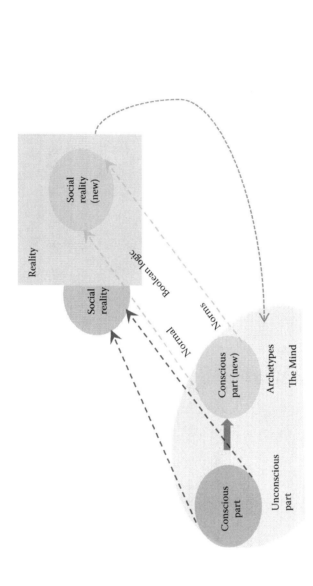

FIGURE 2.3 Eventually, the amount of perceptions that resides outside social reality and are being taken up by the unconscious part of the mind becomes so overwhelming and the increasingly unworldly view by the zeitgeist needs to be replaced with a new view of reality; triggering changes in norms, normality and what is being perceived as taboo or not. This forms a new zeitgeist better tuned to a changed reality.

CONCLUSIONS

That the perception of reality is constrained is well acknowledged. One quickly notes that regardless of academic discipline many features appear almost identical, but given different labelling and the many jargons in describing the functionality of rational thinking, an alignment of the various theories has to date not been feasible.

However, there is little common ground in *how* these blinkers are imposed on humans, what becomes norms, zeitgeist and taboo vary widely over time and culture but it is possible to formulate how they operate and with extracts of data from a *Big Data* approach define its contents and trends over time. But looking at the conscious part of the mind alone will not provide the holistic view of how the mind operates, the unconscious part needs to be understood as well and, in particular, how they interact. Figures 2.1 to 2.3 highlight a suite of depictions demonstrating the functionality of the holistic mind that aptly prompts Chapter 3, which introduces the mechanisms of the unconscious, and Chapter 4, which outlines *The Meta-Model of the Mind*.

ENDNOTES

1. Boole, G. 2003. *An Investigation of the Laws of Thought: On Which are Founded the Mathematical Theories of Logic and Probabilities*. New York: Kessinger Legacy Reprints.
2. Wesley, SP; Nola, JM; Cialdini, RB; Goldstein, NJ; Griskevicius, V. May 2007. The constructive, destructive, and reconstructive power of social norms. *Psychological Science* 18 (5): 429–434. http://assets.csom.umn.edu/assets/118375.pdf (accessed 1 January 2017).
3. Bicchieri, C. 2006. *The Grammar of Society: The Nature and Dynamics of Social Norms*. New York: Cambridge University Press, Chapter 1.
4. Durkheim, E. 2013. *The Rules of Sociological Method and Selected Texts on Sociology and its Method*. New York: Free Press.

5. Shaffer, LS. 2006. Durkheim's aphorism, the justification hypothesis, and the nature of social facts. *Sociological Viewpoints* Fall Issue: 57–70. https://www.questia.com/library/journal/1P3-1639680671/durkheim-s-aphorism-the-justification-hypothesis (accessed 1 January 2017).

6. Bicchieri, C. 2006. *The Grammar of Society: The Nature and Dynamics of Social Norms.* Cambridge, United Kingdom: Cambridge University Press, Chapter 6.

7. Magee, GA. 2011. Zeitgeist, *The Hegel Dictionary.* London, United Kingdom: Continuum International Publishing Group, p. 262.

8. von Franz, M.-L. 1999. *Archetypal Dimensions of the Psyche.* Boston, MA: Shambala Publications Inc., pp. 263–285.

9. ibid.

10. *Encyclopædia Britannica Online.* 2012. Taboo. Encyclopædia Britannica Inc. http://global.britannica.com/topic/taboo-sociology (accessed 1 January 2017).

11. Tetlock, PE; Orie, KV; Elson, B; Green, MC; Lerner, JS. 2000. The psychology of the unthinkable: Taboo trade-offs, forbidden base rates, and heretical counterfactuals. *Journal of Personality and Social Psychology* 78 (5): 853–870. http://www.ncbi.nlm.nih.gov/pubmed/10821194 (accessed 1 January 2017).

12. Cialdini, RB; Goldstein, NJ. 2004. Social influence: Compliance and conformity. *Annual Review of Psychology* 55: 591–621. http://www2.psych.ubc.ca/~schaller/Psyc591Readings/CialdiniGoldstein2004.pdf (accessed 1 January 2017).

13. Turner, ME; Pratkanis, AR. 1998. Twenty-five years of groupthink theory and research: Lessons from the evaluation of a theory. *Organizational Behavior and Human Decision Processes* 73: 105–115. http://www.soc.ucsb.edu/faculty/friedkin/Syllabi/Soc147/Week5Req1Reading.pdf (accessed 1 January 2017).

14. Janis, IL. 1982. *Groupthink: Psychological Studies of Policy Decisions and Fiascoes.* Boston, MA: Cengage Learning, 2nd edition, Chapters 8 and 9.

15. Kihlstrom, JF. 2010. *Unconscious Processes.* University of California, Berkeley. http://socrates.berkeley.edu/~kihlstrm/OxHBCogPsych2010.htm (accessed 1 January 2017).

16. Freud, A. 1992. *The Ego and the Mechanisms of Defense*. London: Karnac Books.
17. Schwartz, C. June 24, 2015. Tell it about your mother – Can brain-scanning help save Freudian psychoanalysis. *New York Times*. http://www.nytimes.com/2015/06/28/magazine/tell-it-about-your-mother.html?_r=0 (retrieved 1 January 2017).
18. Garssen, B. December 2007. Repression: Finding our way in the maze of concepts. *Journal of Behavioral Medicine* 30 (6): 471–481. http://www.ncbi.nlm.nih.gov/pmc/articles/PMC2080858/pdf/10865_2007_Article_9122.pdf (retrieved 1 January 2017).
19. Simon, HA. 1957. A behavioral model of rational choice. In *Models of Man, Social and Rational: Mathematical Essays on Rational Human Behavior in a Social Setting*. New York: Wiley, 1st edition.
20. Kahneman, D. September 2003. A perspective on judgment and choice: Mapping bounded rationality. *The American Psychologist* 58 (9):697–720.
21. Foucault, M. 1982. *The Archaeology of Knowledge*. New York: Pantheon Books.
22. Kendall, G; Wickham, GM. 1992. *Using Foucault's Methods (Introducing Qualitative Methods series)*. London: Sage Publications, p. 42.
23. Kahneman, D; Tversky, A. 2000 *Choices, Values and Frames*. New York: Cambridge University Press.

BIBLIOGRAPHY

Bicchieri, C. 2006. *The Grammar of Society: The Nature and Dynamics of Social Norms*. New York: Cambridge University Press.

Boole, G. 2003. *An Investigation of the Laws of Thought: On Which are Founded the Mathematical Theories of Logic and Probabilities*. New York: Kessinger Legacy Reprints.

Cialdini, RB; Goldstein, NJ. 2004. Social influence: Compliance and conformity. *Annual Review of Psychology* 55: 591–621. http://www2.psych.ubc.ca/~schaller/Psyc591Readings/CialdiniGoldstein2004.pdf (accessed 1 January 2017).

Durkheim, E. 2013. *The Rules of Sociological Method and Selected Texts on Sociology and its Method*. New York: Free Press.

Encyclopædia Britannica Online. 2012. Taboo. *Encyclopædia Britannica Inc.* http://global.britannica.com/topic/taboo-sociology (accessed 1 January 2017).

Foucault, M. 1982. *The Archaeology of Knowledge.* New York: Pantheon Books.

von Franz, M.-L. 1999. *Archetypal Dimensions of the Psyche.* Boston, MA: Shambala Publications Inc., pp. 263–285.

Freud, A. 1992. *The Ego and the Mechanisms of Defense.* London: Karnac Books.

Garssen, B. December 2007. Repression: Finding our way in the maze of concepts. *Journal of Behavioral Medicine* 30 (6). http://www.ncbi.nlm.nih.gov/pmc/articles/PMC2080858/pdf/10865_2007_Article_9122.pdf (retrieved 1 January 2017).

Janis, IL. 1982. *Groupthink: Psychological Studies of Policy Decisions and Fiascoes.* Boston, MA: Cengage Learning, 2nd edition, Chapters 8 and 9.

Kahneman, D; Tversky, A. 2000. *Choices, Values and Frames.* New York: Cambridge University Press.

Kahneman, D. September 2003. A perspective on judgment and choice: Mapping bounded rationality. *The American Psychologist* 58 (9): 697–720.

Kendall, G; Wickham, GM. 1999. *Using Foucault's Methods (Introducing Qualitative Methods Series).* London: Sage Publications.

Kihlstrom, JF. 2010. Unconscious Processes. University of California, Berkeley. http://socrates.berkeley.edu/~kihlstrm/OxHBCogPsych2010.htm (accessed 1 January 2017).

Magee, GA. 2011. *Zeitgeist, the Hegel Dictionary.* London, United Kingdom: Continuum International Publishing Group.

Schwartz, C. June 24, 2015. Tell it about your mother – Can brain-scanning help save Freudian psychoanalysis. *New York Times.* http://www.nytimes.com/2015/06/28/magazine/tell-it-about-your-mother.html?_r=0 (retrieved 1 January 2017).

Shaffer, LS. 2006. Durkheim's aphorism, the justification hypothesis, and the nature of social facts. *Sociological Viewpoints* Fall Issue: 57–70 https://www.questia.com/library/journal/1P3-1639680671/durkheim-s-aphorism-the-justification-hypothesis (accessed 1 January 2017).

Simon, HA. 1957. A behavioral model of rational choice. In *Models of Man, Social and Rational: Mathematical Essays on Rational Human Behavior in a Social Setting.* New York: Wiley, 1st edition.

Tetlock, PE; Orie, KV; Elson, B; Green, MC; Lerner, JS. 2002. The psychology of the unthinkable: Taboo trade-offs, forbidden base rates, and heretical counterfactuals. *Journal of Personality and Social Psychology* 78 (5): 853–870. http://www.ncbi.nlm. nih.gov/pubmed/10821194 (accessed 1 January 2017).

Turner, ME; Pratkanis, AR. 1998. Twenty-five years of groupthink theory and research: Lessons from the evaluation of a theory. *Organizational Behavior and Human Decision Processes* 73: 105–115. http://www.soc.ucsb.edu/faculty/friedkin/Syllabi/ Soc147/Week5Req1Reading.pdf (accessed 1 January 2017).

Wesley, SP; Nola, JM; Cialdini, RB; Goldstein, NJ; Griskevicius, V. May 2007. The constructive, destructive, and reconstructive power of social norms. *Psychological Science* 18 (5): 429–434 http://assets.csom.umn.edu/assets/118375.pdf (accessed 1 January 2017).

CHAPTER 3

How Does the Unconscious Part of the Mind Operate?

The center that I cannot find is known to my unconscious mind.

The Labyrinth by W.H. Auden (1907–1973)

With the recent advances in psychology and neuroscience, the current view of the function of the mind has come closer to the pioneering theories of Freud and Jung, and a suite of evidence has proved the existence of an unconscious with its own logic. Research also suggests that the conscious and unconscious parts can have seemingly conflicting objectives and that these conflicts can result in what in the Freudian era were referred to as neurosis and hysteria but today carries more defined labels under the clusters of *neurotic disorders* and *somatic symptom disorders*. Regardless of this wandering nomenclature, the manifestations are clearly recognisable in the earliest medical descriptions through the various displays of abnormal behaviour and diffuse characteristics, with anxiety and emotional excesses being the main exhibits.[1,2]

HOW DOES THE UNCONSCIOUS EXPRESS ITSELF?

Symbols

For both Carl Gustav Jung and Sigmund Freud, it was clear that the unconscious communicates through symbols, and dreams were typically reviewed for symbols that were then interpreted to understand the contents of the unconscious. That symbols and symbolic language hold a direct access to the unconscious is a claim that has since been supported by a number of empirical tests.[3]

A *symbol* is an object or concept that defines an idea, a process or a physical entity. The word itself is derived from the ancient Greek *symbolon* meaning token or watch word. Symbols come in many forms, images being the most obvious representation, however there are also word symbols as well as symbolic rituals and activities, predominantly religious ones.[4] The use of symbols seems to be an exclusively human form of communication and goes back to the earliest days of mankind with the earliest identified symbols found in caves in Europe some 30,000 years ago. Given that symbols impact most, if not all, aspects of human behaviour, studies of symbols are generally included in a plethora of academic disciplines such as psychology, culture, religion, art, architecture and literature.[5]

It is, however, necessary to distinguish between symbols and signs. Both symbols and signs aim to convey meaning, where they differ is in terms of abstraction. A sign stands for something known, for example, a traffic stop sign means stop and nothing else. Thus, the meaning of a sign is absolutely known, resting on conventions or collectively made conscious agreements.[6]

In contrast, a symbol in addition to its apparent superficial meaning also holds an abstract meaning, which usually has little to do with the dictionary explanation of the symbol itself. It is this abstract meaning that links to the unconscious. Symbols, therefore, provide a unique feature in that they directly access the unconscious, on an otherwise not approachable path through conscious means. Through extensive testing and research in analytical psychology, it is generally agreed that people do absorb and relate to symbols at the unconscious level.[7]

The relationship between symbols and their influence on behaviour has been studied both from the individual and collective perspective, and extensively so in the area of advertising. Advertisers and researchers of consumer patterns have long deployed symbols that unconsciously relate to motherly comfort, masculine strength and reliability, prestige, sexual appeal or any other typical positive attributes they want their product associated with. Also in individual patient cases with behaviour considered abnormal or neurotic, it has been demonstrated that such behaviour can be triggered or accentuated through exposure to symbols representing the root of the mental ailments.

For example, if someone has suppressed the memories of being abused as a child by a person regularly dressed in red clothing, in adult life, just the sight of the colour red can trigger anxiety in that person due to the symbolic meaning that red has come to represent as pain and abuse.

However, there seems to be no single generally accepted approach on how to identify symbols that can be understood by their culturally transmitted meanings. Whether something should be interpreted as a symbol or a sign ultimately depends on the predominant attitude of society. Some symbols become widely used in a given societal context; those symbols and their abstract meanings often become so culturally pervasive that they evolve into a lasting part of both the culture and the way the culture conveys itself. But if the symbolic meaning transcends and in effect replaces the word's literal meaning and can be fully explained in a way we perceive as entirely rational or its abstract charge loses its power and becomes unknown, then it relapses into a sign. There is also the case of 'dead' symbols, which lost their power to influence; examples are the various symbolic terms that used to represent the ancient mythology of Greek and Roman gods, and most of this vocabulary would now fail to resonate in any manner.[8]

Metaphors

A *metaphor* is a literary arrangement designed to assign meaning to a situation or object by drawing comparison or associating to something generally completely unrelated in its dictionary meaning to what is being described.

Through shared references, the non-literal meaning of the metaphor is understood by the general public.[9] The word metaphor stems from the Greek *metaphorá* translating as transfer or to carry over.[10] Metaphors seem to be universal and exist in all known languages but are not exclusively used in linguistics. They are found in other forms of communication, for example, in art work and other graphic representations and even music. Metaphors also exist in extended forms such as anecdotes and parables that provide a broader illustration of the conveyed message. The American linguist George Lakoff, the leading academic in metaphor studies, has shown that emotions are often described as metaphors, such as anger being symbolically conceptualised as heat, wild animals etc.[11]

Examples of metaphors include the following:

- Life is a journey.
- Time is money.

These highlight a common design format of metaphors with a source and a target, with the source (here *journey* and *money*) representing the concepts used to describe or resemble and the target what is being described and understood through the metaphor (*life* and *time*).

The common themes that metaphors typically consist of include the following:

- The human body (and its characteristics)
- Health and illness
- Animals
- Plants
- Buildings and construction
- Machines and tools
- Games and sports

- Money and economic transactions
- Heat and cold
- Light and darkness
- Forces
- Movement and direction[12]

Metaphors are usually applied to describe the following targets:

- Emotions
- Desire
- Morality
- Thought
- Society/nation
- Economy
- Human relationship
- Communication
- Time
- Life and death
- Religion
- Events and actions[13]

As highlighted above, metaphors typically apply tangibles to describe intangibles, describing mental states, groups and processes and personal experiences.

Metaphors are part of *idioms*, or expressions with figurative meaning – in essence any grouping of words whose meaning becomes different from that of each word looked at individually – which also includes metonymy, proverbs, parables and analogues. The English language alone contains at least 25,000 idiomatic expressions.[14]

The key question for groups that routinely deploy metaphors, such as in advertisement and politics, is what is it that makes a metaphor theme work and why some fall flat? They seem to be trending over time, something which reading a dated newspaper or magazine clearly displays; buzzwords and expressions that were in vogue yesteryear sounds out of date and sometimes even outright silly in a current setting. As noted earlier, behind a metaphor one can trace categorisations and through tracking these, it allows for a facilitated digestion of their trending in popularity over time. Through such insights, metaphors bring with it an added element of knowledge; using something contemporary, known to provide perspective to something less well understood. What makes for a 'successful' metaphor is in the matching between similarity and difference. Is it so that the 'successful' proximities in the metaphors highlights what is important in a society, although it might not be consciously obvious but only resonates with the unconscious, hence the themes of metaphors that work are therefore sometimes not that self-explanatory. This would explain the often-failed advertising and political campaigns as their insights do not incorporate the unconscious undercurrents that reverberate with the general population.

As metaphors seem to be founded on the distinctions between similarities (symmetry) and differences (asymmetry), does it mean that they are hiding the categorisations that binds the conscious and unconscious parts together? If we need that connect between the conscious and the unconscious, as we have seen that not all perceptions can be absorbed by the conscious due to capacity concerns and normative restrictions brought in by the reigning zeitgeist with some ending up in the unconscious, metaphors could be one of the connectors through its symbolic categorisation structures. The French philosopher Paul Ricoeur (1913–2005) who wrote some of the most famous works on metaphors pointed to the presence of *absurdity* as a decisive factor as to why some metaphors work and others do not. One might assume, although it was never directly spelled out by Ricoeur, that absurdity was defined in contrast to the contemporary, in essence being a content of the repressed.[15]

A metaphor works to describe what is confusing by providing comparisons and proximity to what is familiar, in that sense it highlights the distinction where the boundary between knowledge and what is new goes. It also reveals what the suitable proximity will be as well as the principles the metaphors put between what is known and unknown in relation to each other. Therefore, the interpretation of metaphors becomes a key tool to understand the current subjective perception of reality, it gives insight to the themes of the prevailing zeitgeist.

One can view metaphors and other types of symbolic language as the unconscious representations in a given text. Freud saw it as the repressed unconscious desires, ideas and fantasies externalising itself organised through common themes and that these start to influence a society's perception of reality. Symbolic language is hence a deliberate act from the unconscious in order to allow for repressed material to find an appropriate outlet to mitigate the risk for mental upsets. In fictional text, the symbolic language typically only consists of a few percentages of the total text mass, and by isolating these and analysing them, the unconscious themes and emotions can be extracted. In this unconscious language, taboo themes can appear in the form of covert symbolism, and by deciphering these, an understanding of whether some taboos are about to disappear from the social reality can be gauged.

THE PROCESS OF ABSTRACTIONS AND GENERALISATIONS

That the unconscious applies categories to sort perceptions has been studied from many perspectives, among them empirical reviews of cultural stereotypes of people by ethnicity, age and gender, indicating that this is done automatically and unconsciously, the degree of categorisation is also not moderated through an individual's level of consciously explicit stereotyping. It seems innate. Hence, the associative aspect of the unconscious is girded through these categories.[16]

These implicit cognitions are the sorting structures from the unconscious on knowledge, perceptions and memory that influences behaviour and as such operates without our awareness thereof. The manifestations of the unconscious structures form generalisations and abstractions, to which stereotypes and metaphors belong. The description of archetypes by Jung also fits these characteristics by drawing associated information towards them and thereby forming strawmen out of the contents. An abstraction becomes a process that acts as a super-category denominator for all subordinate concepts and connects these into unified symmetry.

The formation of conceptual abstractions can be arranged through the filtering of information, selecting an atom of common denominators drawn into a concept of observable phenomena. Abstraction is the opposite to specification, or asymmetry, which is the breaking down of general ideas into distinct particulars.

An abstraction is thereby a process of compression, mapping multitudes of different data into a single data point, with properties that somehow are shared through similarities by each piece of associated data.

Abstraction becomes an exercise in renouncing certain properties in an object or action and puts the emphasis on others and how that categorisation comes about depends on the unconscious structures. The abstraction process works as the symmetry forming of objects and actions. These generalisation patterns become the activated magnets, which the unconscious structures draw behaviour towards.[17]

THE THINKING OF IGNACIO MATTE BLANCO

Ignacio Matte Blanco (1908–1995) was a Chilean psychoanalyst greatly influenced by Freud's views of the unconscious. He elaborated on Freud's five definitions of the unconscious:

1. *Timelessness*: Which among others means that no causality applies, so that A -> B and B -> A can apply at the same time.

2. *Displacement*: Allows for two objects appearing in the same space to be placed in the same class and thus becoming identical.

3. *Condensation*: Memory reflections from separate points in time can be united to an aggregated picture. Time is therefore dissolved and can be bound together by a shared class.

4. *Replacement of external reality by internal reality*: In other words fantasy can be perceived as reality.

5. *Absence of mutual contradiction*: The wishes 'I want to be big' and 'I want to be small' can exist at the same time, as such negations no longer apply.[18]

Out of these five definitions, Matte Blanco, like Freud, meant that the most important characteristic of the unconscious is its aspiration towards symmetry, which he viewed one could identify in dreams where one typically finds deviations from the rational logic that segregates concepts such as space, time, causality, direction and the absence of contradiction and negation. He concluded that if these characteristics are permanent, they must follow some sort of rules as otherwise there would be chaos. Matte Blanco eventually developed a suite of axioms for the structure of the unconscious mind, pointing out the anomalies from Boolean logic. In his most famous work, *The Unconscious as Infinite Sets: An Essay in Bi-logic*, from 1975, he introduces the two main principles of the unconscious logical framework; *the principle of generalisation* and *the principle of symmetry*. With generalisation, Matte Blanco meant that the unconscious logic does not recognise individuals per se but only relate to them as members of classes and sub-classes. With symmetry, he referred to any relation as identical, such that Adam can be the father of Ben but at the same time Ben is also the father of Adam. And while the principle of generalisation does not itself contravene Boolean logic, in amalgamation with the principle of symmetry, discontinuity is introduced, given that relationships are treated as reversible. So, while asymmetric logic distinguishes separate objects from each other through the relationship between them, symmetric logic regards relations as fully reversible between the objects. Matte Blanco pointed to the concept of *project identification* where feelings that cannot be allowed into the conscious are projected onto another person or group that then becomes the embodiment of these particular feelings. Unconsciously that person then behaves in such a way that this repressed and projected feeling is triggered by the other person. Matte Blanco saw this as a process of symmetrisation.[19]

For Matte Blanco, the unconscious is marked by symmetry where sameness is preferred and the unconscious processes work with classificatory activity; they seek out the similarities between things whereas for the conscious distinguishing differences, or asymmetry, is the focus. These two logic systems work in a mixture in what he termed the *bi-logic thinking system*, which he divided into five strata. And it is at moments when symmetric thinking breaks the realms of the traditional logic of the conscious part of the mind that our thought process shifts to unconscious thinking. In each of these five strata, which are operating in a transcending manner, there are particular combinations of symmetrical logic and asymmetrical logic morphed into bi-logic. The five strata operate from the full spectra over the conscious through to the unconscious.

Stratum One

This is characterised by solely operating through the rules of Boolean logic (or similar), in which conscious awareness delineates separate objects and responds in a rational manner.

Stratum Two

It is defined by largely rational reasoning interspersed with symmetric thinking. As an example, a person directing an emotion towards another person, whether that of hate, love or something else, associates all the features of the group the person belongs to, such as gender, race, religion, age or another class differentiator, to that person, while at the same time also being able to recognise the individual aspects distinguishing this person from the attributed class.

Stratum Three

In this stratum, different classes are identified and separate objects of a class will always be regarded as representing the whole class. However, symmetric thinking has become the dominating feature in contrast to stratum two with elements of asymmetrical logic in juxtaposition. Thus, in stratum three, generalisation or stereotyping would start to appear as a standard element in that a single object can be seen representing the whole class. Also, emotions without boundaries begin appearing, examples such as outright rage instead of anger, directed towards the whole class rather than the individual object. An example of generalisation would be a mother who falls under the class of maternal, in other words, all mothers are seen as maternal, regardless or not whether they actually are, identical in sharing a suite of characteristics that defines that class. These characteristics are used to develop the content of an emotion or descriptive attribute (or in a grammatical perspective an adjective) for what is considered maternal. The class itself, in this example maternal, is defined in terms of asymmetric logic differing from that of being paternal but the characteristics inside of each class and all the elements contained in it, such as each individual mother, are ruled by the principle of symmetry.

Stratum Four

Within stratum four, symmetrisation is further extended with ever wider classes and less elements of rational thinking. As per the example above, maternal is grouped with paternal, childlike and so on, forming a higher class of being human. Schizophrenia can occur at this stratum.

A rage occurring would be directed towards everything and anything, no longer aimed at only object and class, like that of a child in a tantrum, kicking and screaming indiscriminately.

From the psychoanalytical perspective, stratum four represents the unknown unconscious rather than what is being repressed in the unconscious and the characteristics that Freud pointed out, timelessness, displacement, condensation, replacement of external reality by internal reality and absence of mutual contradiction applies in full force.

Stratum Five

This stratum is defined as exclusive symmetric thinking that aspires towards unity, in other words *one*, so that everything equals everything. Rational thinking and Boolean logic breaks down into delusion.

In effect, rational thinking will be operating at the surface while the stratum containing ever more symmetrical thinking follows at deeper levels until stratum five, when total symmetry is reached, so that objects that for the conscious mind are separated are united through classes and eventually brought together in unity given the absence of the law of contradiction. Displacement and condensation allow for equal execution of objects in the same space. Timelessness removes differences in time. Ideas, mental phenomena and reality are fully exchangeable.

According to Matte Blanco, normal development involved the ability to being able to transcend and distinguish between these five strata but in an abnormal state, this continuity of differentiation between the strata becomes confused with one's thinking flicking randomly between strata.

Obviously, the 'normal' mode of thinking will be at stratum one and two. The continuum from the conscious to the unconscious part of the mind can be seen as a move from infinity towards one, as asymmetric thinking goes towards an infinite split of objects and symmetric thinking aspires towards a unity of everything and everyone into one. The understanding of unconscious thinking with its classes working to symmetrise individual objects and what these particular classes are can be understood through studying and sorting the symbolic language through which the unconscious expresses itself, whether through idioms, Freudian slips, inappropriate jokes, absurdities and poetic language.[20]

CONCLUSIONS

With the most up-to-date research confirming the unconscious having a capacity vastly larger than that of the conscious, the perceptions and information we absorb will not all be stored in the conscious part of the mind but most of it in the unconscious. These unconsciously registered perceptions are organised and structured in classes; Jung referred to these as archetypes, other research labels them as implicit stereotypes. Ignacio Matte Blanco expanded on Freud's postulates of the unconscious and provided a mathematical expression of the distinction between conscious and unconscious thinking, in essence a continuum move from infinity towards unity. This can be formulated through relaxing the traditional rules that applies for Boolean logic and the introduction of another suite of rules that allows for this aspiration towards symmetry among objects. However, Matte Blanco remained unclear on the actual interaction and transitions between the stratum he outlined and why there were supposed to be five of them, rather than any other number, as it appears arbitrarily chosen. Although many of the tenets of his theory appear promising and broadly conform with current findings in neuroscience, it can only serve as one input in attempting to develop a machine-generated replication of human thinking.

Chapter 4 will outline a meta-model on how the conscious and unconscious parts, making up the human mind, interact through a set protocol and describe it through formulating mechanistic rules.

ENDNOTES

1. Schwartz, C. June 24, 2015. Tell it about your mother – Can brain-scanning help save Freudian psychoanalysis. *New York Times.* http://www.nytimes.com/2015/06/28/magazine/tell-it-about-your-mother.html?_r=0 (retrieved 1 January 2017).
2. McGowan, K. April 2014. The second coming of Sigmund Freud. *Discover Magazine.* http://discovermagazine.com/2014/april/14-the-second-coming-of-sigmund-freud (retrieved 1 January 2017).
3. Jung, CG. 1981. Civilization in transition. In *The Collected Works of C.G. Jung,* translated by R. F. C. Hull. Princeton, NJ: Princeton University Press, Vol. 10, 2nd edition, par. 18.
4. Campbell, J. 2002. Flight of the wild gander: Explorations in the mythological dimension – Selected essays, 1944–1968. In *The Collected Works of Joseph Campbell.* Novato, CA: HarperPerennial.
5. Morrison, R. 2011. New method of identifying archetypal symbols and their associated meanings. *European Journal of Social Sciences* 27.
6. Bruce-Mitford, M. 2008. *Signs and Symbols.* London, United Kingdom; Dorling Kindersley Limited.
7. Morrison, R. 2011. New method of identifying archetypal symbols and their associated meanings. *European Journal of Social Sciences* 27 (2011).
8. Ibid.
9. Lakoff, GP; Johnson, ML. 2003. *Metaphors We Live By.* Chicago: University of Chicago Press, 2nd edition.
10. Liddell, HG; Scott, R. (edited by). μεταφορά. *in A Greek–English Lexicon.* Perseus. http://www.perseus.tufts.edu/hopper/ (retrieved 1 January 2017).

11. Lakoff, GP; Johnson, ML. 2003. *Metaphors We Live By*. Chicago: University of Chicago Press, 2nd edition.
12. Kovecses, Z. 2010. *Metaphor: A Practical Introduction*. New York: Oxford University Press, 2nd edition.
13. Ibid.
14. Jackendoff, RS. 1996. *The Architecture of the Language Faculty*. Cambridge, MA: MIT Press.
15. Ricoeur, P. 2003. *The Rule of Metaphor: Multi-disciplinary Studies of the Creation of Meaning in Language*. London: Routledge Classics, Taylor & Francis Group, University of Toronto Romance Series.
16. Devine, PG. 1989. Stereotypes and prejudice: Their automatic and controlled components. *Journal of Personality and Social Psychology* 56 (1): 5–18. https://www.uni-muenster.de/imperia/md/content/psyifp/aeechterhoff/sommersemester2012/sozialekognition/devine_automcontrprejudice_jpsp1989.pdf (retrieved 1 January 2017).
17. Perruchet, P; Gallego, J; Savy, I. October 1990. A critical reappraisal of the evidence for unconscious abstraction of deterministic rules in complex experimental situations. *Cognitive Psychology* 22 (4): 493–516.
18. Rayner, E. 1995. *Unconscious Logic: An Introduction to Matte Blanco's Bi-Logic and Its Uses*. London: Routledge, The New Library of Psychoanalysis.
19. Ibid.
20. Matte Blanco, I. 1988. *Thinking, Feeling, and Being: Clinical Reflections on the Fundamental Antinomy of Human Beings and World*. London: Routledge, The New Library of Psychoanalysis, Vol. 5, 1st edition.

BIBLIOGRAPHY

Liddell, HG; Scott, R. (edited by). μεταφορά. *in A Greek–English Lexicon*. Perseus. http://www.perseus.tufts.edu/hopper/ (retrieved 1 January 2017).
Bruce-Mitford, M. 2008. *Signs and Symbols*. London, United Kingdom: Dorling Kindersley Limited.

Campbell, J. 2002. Flight of the wild gander: Explorations in the mythological dimension – Selected essays, 1944–1968. In *The Collected Works of Joseph Campbell*. Novato, CA: HarperPerennial.

Devine, PG. 1989. Stereotypes and prejudice: Their automatic and controlled components. *Journal of Personality and Social Psychology* 56 (1). https://www.uni-muenster.de/imperia/md/content/psyifp/aeechterhoff/sommersemester2012/sozialekognition/devine_automcontrprejudice_jpsp1989.pdf (retrieved 1 January 2017).

Jackendoff, RS. 1996. *The Architecture of the Language Faculty.* Cambridge, MA: MIT Press.

Jung, CG. 1981. Civilization in transition. In *The Collected Works of C.G. Jung*, translated by RFC Hull. Princeton, NJ: Princeton University Press, Vol. 10, 2nd edition.

Kovecses, Z. 2010. *Metaphor: A Practical Introduction* (New York: Oxford University Press, 2nd edition).

Lakoff, GP; Johnson, ML. 2003. *Metaphors We Live By*. Chicago: University of Chicago Press, 2nd edition.

Matte Blanco, I. 1988. *Thinking, Feeling, and Being: Clinical Reflections on the Fundamental Antinomy of Human Beings and World*. London: Routledge, The New Library of Psychoanalysis, Vol. 5, 1st edition.

McGowan, K. April 2014. The second coming of Sigmund Freud. *Discover Magazine.* http://discovermagazine.com/2014/april/14-the-second-coming-of-sigmund-freud (retrieved 1 January 2017).

Morrison, R. 2011. New method of identifying archetypal symbols and their associated meanings. *European Journal of Social Sciences* 27.

Perruchet, P; Gallego, J; Savy, I. October 1990. A critical reappraisal of the evidence for unconscious abstraction of deterministic rules in complex experimental situations. *Cognitive Psychology,* 22, (4).

Rayner, E. 1995. *Unconscious Logic: An Introduction to Matte Blanco's Bi-Logic and Its Uses.* London: Routledge, The New Library of Psychoanalysis.

Ricoeur, P. 2003. *The Rule of Metaphor: Multi-disciplinary Studies of the Creation of Meaning in Language*. London: Routledge Classics, Taylor & Francis Group, University of Toronto Romance Series.

Schwartz, C. June 24, 2015. Tell it about your mother – Can brain-scanning help save Freudian psychoanalysis. *New York Times*. http://www.nytimes.com/2015/06/28/magazine/tell-it-about-your-mother.html?_r=0 (retrieved 1 January 2017).

The Mind

The Meta-Model

Reality exists in the human mind, and nowhere else.

Nineteen Eighty-Four by George Orwell (1903–1950)

In order to design a meta-model of the mind, the interaction between the conscious part with its deductive logical reasoning with a truncated view of reality and the unconscious part with its own set of logic becomes the key challenge. How are they amalgamated, and how does the unconscious logic intermittently supersede rationality; what are the triggers, and can they be prognosticated? As a starting point and in accordance with Matte Blanco's bi-logic framework, they seem to be interspersed in various degrees through a continuum, extending to each extremes with strict rationality at one end and the aspiration towards levelling everything into unity at the other.

One hypothesis that describes this interaction has been introduced by the Swedish psychologist Jurgen Reeder, who states that conscious thinking can be depicted as a horizontal unbroken line representing the continuous absorption of perceptions. The unconscious, however, is not depicted as a parallel line but through separate vertical lines appearing sporadically and intersecting the conscious mind's line. While this horizontal line indicates movement along time with a clear distinction between the past, present and future, the unconscious with its vertical line highlights its timelessness as indicated by Freud et al. The points of intersection are the typical Freudian slips, the funny but inappropriate jokes that go against the reigning social norms, intuitions, sudden emotional outbursts or any other manifestations that can be characterised as irrational. These are the moments when the unconscious supersedes conscious thinking.[1]

To develop a meta-model of human thinking, the first design criterion, therefore, needs to be able to timely and accurately replicate when the irrational thinking emanating from the unconscious will override that of the conscious mind's rational thinking. What is it that suddenly sparks an outburst of unconscious activity?

The second criterion will be to understand what unified classes the unconscious will direct its energy towards. Are these different from the current classes that dictate the zeitgeist, as such can they provide cues and indications of a possible change of the zeitgeist? Is it something or someone as part of the perception in question that causes this symmetric association to classes in the unconscious to trigger?

The third criterion relates to how unconscious thinking will influence the otherwise anticipated rational thinking in decision making. Can one assume that such a decision will be anything but following rational logic and it will be focused around the dominating class within the unconscious, which will come to colour its preferences? As referred above, these interactions when the unconscious overtakes conscious thinking occur only temporarily and in the *short* term, but the meta-model must also be able to establish the mechanics on how they interact in the *long* term. The conscious mind's view of reality is constrained through social norms but the collection of norms, or zeitgeist, changes over time and the repressed perceptions stored in the unconscious influences the contents of the new zeitgeist and its directions. A key criterion of the meta-model is the need to incorporate when such zeitgeist change is likely to take place and its composition as well as what might precede such a change, like whether there is an increased level of injunctions from sudden unconscious manifestations.

To develop the meta-model, we first need to go back to various linguistic and philosophical findings as they hold clues to the juxtaposition between conscious and unconscious thinking.

GOTTLOB FREGE

Gottlob Frege (1848–1925) was a German philosopher best known for his academic work in the philosophy of language and mathematics. One of his major achievement was in presenting a distinction between *Sinn* and *Bedeutung*, usually translated as *sense* and *reference*. This distinction commences with the analysis of the expression 'a = b'. What Frege is trying to seek is a solution to why the expression 'a = a' does not seem to contain any useful information, in contrast to the expression 'a = b' that apparently does so. It cannot be that 'a' and 'b' are in all aspects identical as 'a = b' that would be synonymous with 'a = a'. Thus, there must be a distinction between 'a' and 'b' at the same time as there must exist a similarity. What Frege proposes is that 'a' and 'b' have the same reference (Bedeutung) but different sense (Sinn). By reference, he points to the actual physical object and by sense its conceptual content or the way the expression's reference is given by the expression itself. To highlight this distinction, Frege used the example of the names of the planet Venus, called both 'evening star' and 'morning star', and according to him, both expressions have the same reference as they both refer to the planet Venus, but they have different sense, in that the 'evening star' appears in the evening and the 'morning star' in the morning. In this manner, the expression 'evening star = morning star' is different from the expression 'evening star = evening star'.[2]

But to know whether the expression 'evening star = morning star' is true, we need to know something about the semantic content, that is if they are a reference to the same physical object, the planet Venus. For the expression 'evening star = evening star', however, no knowledge of their semantic content is needed, that the statement is true can be inferred through logical deduction alone. From these conclusions came *Frege's principle* which underlines that in a meaningful sentence, if the lexical parts are taken out, what remains will be the rules of composition. This principle works as a definition of an expression's reference and suggests that an expression as a whole is the sum of its components; therefore a composite expression is something that must evolve top-down so that they are defined by the reference of words and not the other way around.[3]

JACQUES LACAN

Jacques Lacan (1901–1981) was a French psychoanalyst, and something of a Freudian disciple, who presented some innovative theories on structural linguistics; the gist of which can be concluded through one of his most famous quotes '*the unconscious is structured as a language*'.[4]

Much of his work consisted of trying to incorporate Freud's more anecdotal postulates into a structural form. Lacan also took inspiration from the Russian linguist Roman Jakobson (1896–1982), as he presented similarities between the defense mechanisms of *condensation* and *displacement* with that of the figurative language of *metonymy* and *metaphors* by linking them to *proximity* and *similarity*.

To recap on these, displacement aims to redirect one's focus from something considered socially unacceptable to something acceptable, and condensation is the combination of two features into one; one unacceptable and one acceptable to dilute the effect of the unacceptable. Metonymy in its original Greek means 'change of name' and it is as such as it works; '*taking to the bottle*' instead of '*alcoholism*' as an example, the name change is due to the condensation between the words 'alcoholism' and 'bottle', here referenced in its meaning as a container to store alcohol. Metaphors refer to the replacement of meanings through an assumption of similarity, hence a displacement such as referring to a bar or other drinking establishment as a 'waterhole'. In linguistics, proximity and similarity are considered the two fundamental poles along which the human language is developed and it holds noted resemblance with Frege's concepts of reference and sense.[5]

RATIONALITY, IRRATIONALITY AND EMOTIONS

Through neuroscience, as we noted in previous chapters, we now better understand how our behaviour is guided by motives and goals and how these steer what we will focus on in reality and what we will filter out. And it is not always necessarily that we are consciously aware of what these motives really are and therefore deliberately focus on. One example is daydreams which we are conscious about but do not really pay any attention to and vice versa; we can pay a lot of attention to matters we are unaware of, such as the expression has it; *'sleeping on a problem'* in trying to solve a complex problem whose solution often seemingly surprisingly appears through an eureka moment. Irrationality is a part of this and is described as a cognitive activity without inclusion of rationality.[6] Typically, irrationality is assumed to be a decision made in violation with rational inference. The expression of exaggerate emotions is often seen as irrational under the premise that emotions are considered as sources of erroneous assumptions but strictly pointing out what is rational or irrational is often difficult because it is rarely that clear what elements of reality are incorporated in the decision-making process. With the benefit of hindsight many often shake their heads at historical decisions, mocking them as irrational, however with the norms and filters that were applied on reality at the time and following standard logic, these were in fact generally highly rational decisions. Over time, a number of theories have evolved on why irrational behaviour occurs at all, contravening the view of the assumption of the rational man theory:

- There is an actual misunderstanding in what one believes to be one's true goal.

- What provides for rational decision making in *normal* settings leads when equally applied to irrational outcomes in *abnormal* settings.

- Moments of high levels of stress causes distorted perceptions, such as over exaggerating the level of danger and this causes the decision-making process to become irrational.

- Ever-widening discrepancies between zeitgeist and reality eventually leads to the foundation of one's worldview becoming so detached from reality that one's actions become irrational as key facts are blinded out.

- What from the conscious mind's perspective seems irrational, actually is fully optimal decisions as the goals and motives of the unconscious differs.

- Out of social acceptance, individuals are forced to make decisions that go against their best interest and this can over time often lead to distress and neurosis.[7,8]

And it is here that the difference between artificial intelligence and psychoanalysis becomes distinct; while psychoanalysis points to aspects of the irrational in the mind as emanating from the unconscious and trying to understand it, artificial intelligence develops methodologies based on the rational alone. Whereas psychoanalysis deals with the ambiguity where irrational states, such as mixed emotions, play a role in behaviour, a concept that artificial intelligence with its rational operator has difficulties dealing with, and hence, any attempt to resemble human behaviour will stumble. But if artificial intelligence can be deployed to provide a rational understanding of the irrational, the two concepts can be integrated as irrationality does not equal chaos but it has structure and thus rules. As intelligence defined through logic needs specific rules, these by necessity must truncate and exclude certain parts of reality simply to function in a complex environment, something which resonates well with the psychoanalytical concepts of repression. The unconscious hosts this redundant information.

The content of the unconscious can carry emotions, which can be expressed mainly in two ways; when the original emotion is represented consciously but we are unconscious of the source of that emotion, as in the displacement and projection defense mechanisms, and where the emotion itself is denied conscious representation, as in reaction formation, intellectualisation and denial.

A person may be consciously aware of his emotional state, yet unaware of its source, as it is of a repressed unconscious nature. As such, emotions can be unconsciously rooted and this trigger seemingly irrational decisions.

And what are these triggers? Generally, when human drives are violated, for instance, 'dislikes' or even 'hate' remain unconscious and drive stereotypic or aggressive behavior. If so, how are they made conscious? Can people's emotional reactions be driven by unattended emotional pictures and words that might influence decisions without ever being explicitly detected?

THE UNCANNY

The uncanny, or in its original German 'das Unheimliche' (translated as 'opposite to what is familiar') is a psychological effect that was described by Sigmund Freud as an amalgamation of something familiar and unfamiliar in one and as such being experienced as strange, frightening and even bizarre. This obvious contradiction can create cognitive dissonance due to the ambiguity of the combination of an attraction or recognition of the familiar and at the same time an animosity of that unknown component within the familiar. That component of the unknown tends to lead to a rejection of the object or situation. Freud linked the concept of the uncanny with social taboos that can provide both a feeling of fascination and also, and more so, cause distaste and horror. So, what is not allowed by the norms of society and hidden away in the unconscious is, in part, contents of the uncanny and that is what reminds us of our own desired but forbidden drives and instincts. This is so, especially when a familiar object is suddenly seen as containing a component of these repressed desires as this can cause an appearingly irrational sense of guilt. Our projections upon others often contain uncanny material and in studying these, an understanding of the content can be obtained, that is what and whom it is that we blame for our misfortunes. As being part of a taboo, the uncanny changes over time, what has been familiar and known can over time become the unknown and then be capable of generating horror.[9]

GESTALT PSYCHOLOGY

Gestalt psychology developed in the early twentieth century with Germany as its main base; *gestalt* being the German word for 'shape' or 'form', as an attempt to pin down and understand the laws that determine human's ability to make sense of the perceptions it absorbs. The main tenet of gestalt psychology is that the mind through structures and principles seeks to form holistic perspectives out of independent parts. One of the leading gestalt psychologists, the German Kurt Koffka (1886–1941), stated that when the human mind perceives a gestalt, '*The whole is other than the sum of the parts'*, as such the whole has an independent existence. Gestalt psychology has come to be mostly applied on how the mind perceives visual intakes such as seemingly unrelated points and lines into a whole and what the laws are that organise these into forms. Gestalt psychology has its foundation in two main principles:

- *The principle of totality*: A conscious experience must be considered holistically as the design of the human mind is such that each individual component is considered as a part of a greater whole. In essence, it seeks out patterns or structures unconsciously to connect dots and points into a whole.

- *The principle of psychophysical isomorphism*: The assumption that a connection exists between a (conscious) experience and activities in the brain and nervous system.[10,11]

Out of these two principles, the gestalt psychologists formed a suite of laws under the German word *prägnanz*, or pithiness, bringing to the forefront that our sorting of perceptions and experiences is regulated through laws that aim towards regularity, orderliness, symmetry and simplicity.

These gestalt laws can help us predict how the mind will interpret new stimuli given the structures and groupings they provide. Although they were mainly defined and empirically reviewed for visual perceptions, their areas of application also include other types of perceptions, such as language. The key ones include the following:

1. *The law of proximity*: A suite of objects in close proximity are viewed together as a collective group under an assumed association.

2. *The law of similarity*: Among a collection of objects, a grouping is done on the premises of some sort of resemblance to each other.

3. *The law of closure*: Fragment of shapes are interpreted as a whole object as gaps are filled in by the mind.

4. *The law of symmetry*: States that objects that are visually symmetrical to each other tend to be perceived as a unified group. Similar to the law of similarity, this rule suggests that objects that are symmetrical with each other will be more likely to be grouped together than objects not symmetrical with each other.

5. *The law of common fate*: Implies the grouping together of objects that have the same trend of motion are therefore assumed to be on the same path.[12,13]

HOW DO THE UNCONSCIOUS AND CONSCIOUS INTERACT? UNDERSTANDING THE SHORT-TERM VERSUS THE LONG-TERM IMPACTS

The sudden flash of irrationality invalidating rational thinking is not an uncommon feature. They override our 'good' judgment and make us act non-sensically; *'not being oneself'* is a common term to describe such occurrences.

Why do these outbursts occur and can we forecast situations when they are likely to happen?

If we can explain them, we are a step closer towards understanding the functionality of the human mind as decisions made in an emotional state do not comply with the rational man theory, assuming a sound logical argumentation preceding every decision and with emotions sourced from a broader perspective of reality distorting that. First of all, one needs to ascertain whether acting irrational is simply the opposite of rationality;

in other words, decisions not deliberated through considering all inputs and engaging Boolean logic in evaluating and ranking them. However, such negative definition is too wide to be meaningful in projecting human behaviour. Additional precision is needed to be able to pin down triggers and the content of sorts to allow for greater accuracy in prognosticating *when* irrational reactions will activate and *in which* directions they will go.

Freud considered being exposed to the repressed material from the unconscious that lies embedded in objects and features in the social reality, through condensations or other defense mechanisms, as the trigger of seemingly inexplicable irrational outbursts which supersedes rational manifestations. The sudden out-of-proportion anger, even rage, that can arise when being exposed to a scapegoat object, person or group upon which projected repressed content have been attached is the typical phenomena.

Through understanding what the existing taboos and repressed contents are and how they uncannily can slip into social reality as *unknown knowns* both the trigger, timing and contents can be estimated.

A reaction will emerge from the unconscious when an object contains something unknown known that raises the affect level and overwhelms the conscious rational thinking.

This reaction becomes the attractor and key consideration in the decision making, such as if the colour 'red' in its symbolic meaning contains an unknown known which links to a taboo or a repressed theme. This will in a distinct way influence the decision-making process in the preferred choice for a blue car over a red car, despite the red car being cheaper, of better quality and other advantages; thus, parts of the symbolic content of the colour red that is only registered by the unconscious invalidates the rational inference. For the holistic mind, when the unconscious detects that decisions within the social reality does not concur with the mind's overall goals, it overrides them. The device to trigger these reactions comes through symmetry seeking functions, in accordance with Gestalt laws, as they aspire towards unity through symbolism which tends to be unrecognisable to the conscious part of the mind, however, still manages to create a response.

WHEN THE UNCONSCIOUS SHIFTS THE ZEITGEIST BOUNDARIES FOR CONSCIOUS THINKING

But the unconscious also works to over time shift the boundaries of social reality, which changes the content out of which the conscious part of the mind bases its rational decisions through recalibrating the social reality.

When changes in norms and dogmas occur, the directions and themes thereof originate from forces in the unconscious. The reasons why we do see changes in the zeitgeists over time we discussed in the preceding chapters; at some point, certain norms that form our lifestyle fail to reasonably resonate with how reality over time have changed since the outset of that particular zeitgeist.

And by remaining within the perception that the now obsolete zeitgeist provides, it produces the allocation of decisions detrimental for mankind's psychological health in general.

Eventually, a society is so hedged in by taboos that most part of reality must be repressed in order not to be considered a social outcast. Obviously, that leads to an increasing number of psychological ailments, typically neurosis, and as a kind of a psychological life instinct, the alignment of the zeitgeist to better correspond with a changing reality emerges from the unconscious and eventually surfaces in the conscious as newly established norms, taboos and normality, all consolidated into a new zeitgeist. What the content of this new zeitgeist will be can early on be detected through symbolisms and their specific generalised classes stemming from the unconscious. But the view that the waxing and waning of zeitgeist classes should be like a shifting pendulum, expecting the emerging one to take the opposite side of the existing one is simply too simplistic. No consistent historical patterns have proven to provide a robust foundation for that notion. As such, no war cycle theories have to date to deliver any accuracy in forecast, other than pattern adjusted and fitted in hindsight, as the regularity in frequency and duration between time epochs of aggression and peace cannot be described through a simple equation with constant numerals. Rather, it is so that the new zeitgeist rectifies any psychological imbalances caused by the widening discrepancies between social reality and physical reality simply by aligning closer to physical reality rather than by default seeking out opposites, albeit they not always need to diverge.

Adequately manifested throughout history is what precedes and act as a first precursor to a shift in zeitgeist, namely a changing psychological mood is generally expressed in a stagnating and neurotic society, sometimes but not always reflected in an economic and cultural life going nowhere, but always in increased number of the population suffering from psychological ailments. These shifting normative systems when a society stands between zeitgeists creates a twilight zone of sorts, which was labelled *anomie* by Emil Durkheim. In essence, it is a condition where the old suite of norms is being questioned as no longer valid and the new normative framework has yet to be fully endorsed. This creates alienation and increased levels of uncertainty among the population while they are finding their ways, triggering anxiety and psychological disturbances and can be gauged as a sign of a changing zeitgeist.

Such notions are, however, hard to objectively assess as definitions of psychological disorders vary over time; many cases go unreported due to the social stigmas they carry and many are treated through 'self-medications', typically alcohol and drug abuses. Therefore, one must instead identify another proxy measurement technique to:

1. Prognosticate when a zeitgeist change is likely to occur by pinning down trigger levels.

2. Prognosticate the theme of the zeitgeist change in order to understand direction.

3. Develop a time series to estimate trends over time.

4. Draw data from publically available sources to allow for objective verification, as far as possible.

By applying the view of the unconscious as being an attractor of symmetry, as described in Figure 4.1, and by evaluating them versus the classes of the reigning zeitgeist, with diverging classes indicating a looming breaking point translatable as an emerging new zeitgeist, the patterns in the unconscious forms a leading indicator for change. The new classes from the unconscious when entering the conscious part of the mind will at first be seen as unusual and unpopular, though not necessarily being the considered taboos. From the perspective of the trends and fads of the social reality and being embedded in the bi-logical structural arrangements, the new classes will appear as symbolic and emotionally charged language preceded by markers for generalisations. It stands in contrast to the language of the conscious mind that towards the end of a

Mechanisms for generalisations (symmetry striving functions)		
Figurative language	Metaphor	Metonymy
Function	Resemblance of something which it is not	Acts through proximity. One part represents the whole
Gestalt	Law of similarity	Law of proximity and law of closure
Defense mechanism	Condensation, acts for the uncanny	Displacement
Fregian terminology	Sinn-sense	Bedeutung-reference

FIGURE 4.1   A comparisons of the mechanisms that acts for symmetrisation, working as attractors for unconscious categories.

particular zeitgeist draws towards ever-increasing asymmetry, like that of the introduction of a plethora of euphemisms to describe and re-describe a social reality too detached from the actual reality.

SYMBOL ANALYSIS

There are a number of approaches used in capturing symbols because they appear in a variety of contexts, although mostly in text formats. When extracting symbol words as representations of the unconscious, it is pivotal to do so only when the word is used in its figurative meaning rather than its literal meaning, as a lack of filtering will contaminate and distort true signals. Among psychoanalysts, the typical mean of collating unconscious content is through active imagination, in short drawing on self-expressions by stimulating fantasies and imagination through focused concentration, a kind of dreaming with open eyes. The content generally manifests in the form of images and narratives, from which symbols can be identified and noted for analysis. Because the technique is performed individually, its main focus is the therapeutic work on the personal unconscious rather than the collective unconscious, although elements of the collective unconscious will also be represented. To incorporate active imagination or dream analysis as part of a measurement methodology has plenty of constraints:

- The symbolic representations from the collective unconscious must be segregated from the ones harboured in the personal unconscious. To ensure such filtering one needs to be able to gather data from a wide and varied population with regards to gender, age, cognitive ability, culture, race and other characteristics. This is to ensure that one captures the common denominators among these individuals, thus representing the collective unconscious and filtering out the individuals' personal elements.

- There is also the obvious problem with an objective interpretation of these subjective symbol displays in the manner various individuals describe them and ensuring that the interpretations can be standardised to avoid ambiguities.

- Other problem areas include determining how to objectively assess the magnitudes of symbol occurrences, as they will not readily lend themselves to a simple count, and the fact that the updating of data needs to be done with a regular frequency. Biannual or annual measurement points would be too infrequent and would lead to data points too scarce to develop a time series that lends itself to statistical testing.

The preferable measurement approach is to draw the existing norms and classes that defines the zeitgeist, the symbolic language that highlights currents in the unconscious as well as the emotional sentiment that provides directions to the various classes from a public media source through text and symbol analysis. The societal collective sentiment, whether consciously or unconsciously represented, can be objectively measured through a big data approach by extracting data from public media, extended and expanded enough to serve as a meaningful proxy of public opinion. This transparency of the data provides the benefit of verification of results through independent testing. The mechanically set extraction rules to define markers for generalisations and a syllabus for symbol words also allow for verification. A word count along a time line provides the time series that facilitate the trend analysis needed for ongoing assessment. There are, however, still a number of subjective areas that remain challenging to entirely overcome; how to define classes, especially symbolism which can only be interpreted in context, such as, what will the colour 'red' in its symbolic meaning refer to? Is it 'blood' in an aggressive class or does it represent 'love' in a more peaceful class or something else? Only in cluster with other symbol words, sharing either of the classes can its symbolic meaning be ascertained, again this obviously leaves room for subjective interpretation. When selecting data sources, the following criteria need to be considered:

- Focus on the local language media relevant for that specific cultural confinements.

- Incorporate national, regional and global coverage.

- Political views of the publication or authors do not matter as the collective mind's influence spans over political divergence. Instead, it is the context within which divergent views operate that will be affected by the emerging thought patterns, although to a different degree depending on ideology. That might lead to polarisation in terms of political views; however, it is not the political directions per se that are dictated by an emerging zeitgeist. For example, a more authoritarian view can be represented both through communism and fascism; they share the underlying sentiment. Being able to exclude such concerns helps to simplify the cleansing procedure and analysis, something that traditional sentiment analysis struggles with as it needs to consider the contextual polarity. Text sentiment analysis also needs to adjust for ironic statements and linguistic nuances required to be factored into the model to be able to capture the emotional state of the author. Incorporating these factors becomes a challenge for the analysis algorithms, and determining the actual underlying sentiment of the text becomes difficult. This highlights the difference between text analysis and symbol analysis as it does not really matter whether the symbols are used in a positive or negative connotation, in an ironic fashion or not; what matters is the fact they are being used and how often which demonstrates their importance in influencing perceptions.

- Generally, the sources are not broken down by subject, as that from the point of the collective unconscious is unimportant, but the development of a time series is done through a consolidated collation, not segregating or weighting included media sources differently. However, at the outburst of a particular symbol class, it is tracked back to explore the origin of source and subject to determine whether it is possible to establish any early warning system indicator and focus the scope of the analytics.

- Any topic that could be addressed using figurative language are to be included, in particular colloquial contents of culture, fashion, fiction, politics and sports articles, but as a rule not technical or scientific documents as these rarely are described through figurative language. The actual texts are in themselves generally not of interest but through the symbols captured, we are interested in the tone the news is presented or described in, and determining the type of thematic metaphorical language being used. This is a fact long noted by media researchers: that a review of news will provide more than just the facts of the specific tenets it aims to describe. The way the news is framed provides insight into the reigning zeitgeists.

- The media coverage is generally text based but will also include transcription of interviews and speeches. Periodicals and books are not included in the database searches given time lags. While this may pose a limitation in scope, the particular symbol words in fashion are assumed not to be included only in literature but instead to permeate all written communication and public types of broadcasts. Hence, not being able to include these genres should not distort the analysis.

• With regards to double counting, if articles or part of articles are replicated or quoted, these will be counted as many times as occurring in order to gain an understanding of each article's popularity. As part of the framework, circulation or hits per article are not calculated and used to adjust for frequency, as there are technical issues in ensuring comprehensive access to obtaining hits for all articles included in the media sources and cannot be consolidated with circulation numbers of the articles in paper form.

Capture and Cleansing

Once the data sources containing news media have been identified and prepared for extraction of symbol words, capturing and cleansing procedures need to be established to facilitate the development of a time series used for statistical testing. There are four major steps involved:

1. Apply the symbolic language upon the data sources and ensure that filters are employed so that only words or phrases used in symbolic and generalised form are captured, and are excluded when used in their literal meaning.

2. Categorise the captured words by class.

3. Relativise the symbol words through assessing their frequency versus the total text mass and highlight their

relative, rather than absolute, magnitude as the text volumes vary over time, which will distort the trend if using an absolute measurement approach.

4. Conduct in-depth reviews if any spikes are detected in the times series versus set thresholds, to investigate the reason for the spike. Why the sudden drastic increase in use of these symbol words and are they contrasting the existing zeitgeist? If the investigation points to the inclusion of erroneously labelled figurative language not filtered out by existing rules, then write off the findings. This will facilitate calibration of exclusion filters. If the spike is 'legitimate', conduct a review to determine the origins of the spike, in terms of source, topic and geography, with a view to establishing early warning indicators. Contrast the classes of the symbol words with existing trends defining the zeitgeist and determine patterns of divergence.

Once the cleansing is completed and only the symbol words remain, with everything else removed and stripped out, so the symbols then become clearly visible, what then should emerge and be revealed on the aggregate level, overarching any news contents, topic, political view or geographical location, is the language of the unconscious. It is only through the consolidated study of all-encompassing media sources that hidden patterns will appear in the text layers and this generally represents only a small portion of the total text mass. At first, authors and audience are consciously unaware of the unconscious thought patterns they are communicating and receiving through the symbols, which reinforces its influencing power. Eventually it enters the conscious, these symbols become fads and fashions, their frequency increases exponentially and starts to influence behaviour and decision making on a collective scale, that is a new zeitgeist has been established capable of drastic societal change.

In addition to using contextual filters to help rule out the literal meaning of words instead of its symbolic meaning, there are a number of other rules that help in the cleansing of symbol words:

- It does not matter in what tense the figurative language is being presented: past, present or future. It is the time stamp of the article or news piece that matters and dictates to what date in the time series the occurrence of the symbol word is pinned, even if the particular section including the symbolic language related to a previous quotation. The unconscious does not distinguish between tenses.

- Ignore whether a symbolic word is preceded by a negative or presented in an ironic manner reversing its meaning. This does not cancel out or revert its influence; studies in hypnotic technique have demonstrated that the unconscious does not consider negations. The important thing is that the particular symbol words are being used, even if their meaning is reversed.

- Do not differentiate whether the symbolic language is being used to advocate for or against a particular topic. Again, it is the actual increased use of a theme of symbols that highlights what kind of thought patterns are dominating the debate and what type of outcomes are likely, whether financial, political or social.

- Count repetitions of symbols in the same text. If they occur as referrals or quotations from article to article, this gives an indication of the symbols' popularity.

- Ignore whether the symbol words appear as singular or plural. They should be counted as per appearance in each text source.

Relativisation

As the text volumes included in the media sources will vary over time, taking an absolute count approach of the symbol words risks distortion. That is, its volumes can go up as the total volume of text increases but that really will not convey its relative importance if the increase is not equal percentage-wise. Therefore, the number of symbol words as counted per time unit needs to be relativised with the total number of words included in the media sources. As a matter of simplification, linked references in online articles are not included, as they might or might not contain symbol words; but if not appearing in the included articles of that particular time unit will not be counted. The symbol words' relative popularity is not assessed on an article-by-article basis by only counting those which contain symbol words, but on the total number of words that constitutes the included articles of the time unit, whether actually including symbol words or not.

THE COMPONENTS OF A VIRTUAL MIND

There is merit in pairing the concepts of bi-logic, archetypes and generalisations with the dual process theory as they share and overlap a number of features, albeit applying a differing nomenclature in describing them. The benefits of uniting these frameworks are that the extended approach from Freudian and Jungian perspectives provide a juncture to determine and measure the constraints in the rule-based rational thinking, the classes to associative thinking through analysing symbolic language, such as metaphors in common use in public media. Through a big data collation exercise, symbolic language can be extracted, categorised and its frequencies tracked over time to determine trends. With an elongated measurement period, it will allow for determining the existence of any dynamic constants in terms of duration and magnitude, the change of class associative patterns and the establishment of a rule-based mechanism for patterns of human thinking and how the unconscious reasoning intermittently supersede conscious reasoning; what are the triggers and how are they prognosticated. Figures 4.2 and 4.3 depict the shifts of social reality over time.

This chapter has in a prosaic manner outlined the components that constitute the building blocks in the design of a virtual mind:

The conscious part of the mind

- A calculation device for Boolean logic operations representing the algorithms for rational thinking.

- An ongoing feed of perceptions to the Boolean logic device, which is drawn from a big data media source representing a proxy of the collective mindset.

- A filter applied on the perception feed to distinguish the social reality from the physical reality, constructed to highlight the classes allowed for by the zeitgeist.

The unconscious part of the mind

- A calculation device for the unconscious logic based on the principles by Matte Blanco that aspires towards symmetric generalisations.

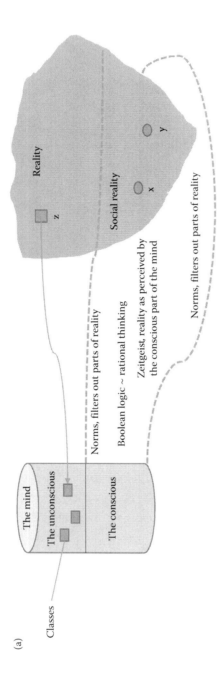

FIGURE 4.2 (a) A depiction of the filtering out of reality. Three objects to consider in a decision making situation; X, Y and Z, where X > Y and Z > X, but with Z existing outside social reality being blinded out, the so-called elephant in the room. From the perspective of the conscious mind, the selection of X represents a rational decision, however holistically it is a suboptimal decision as Z is the largest entity. *(Continued)*

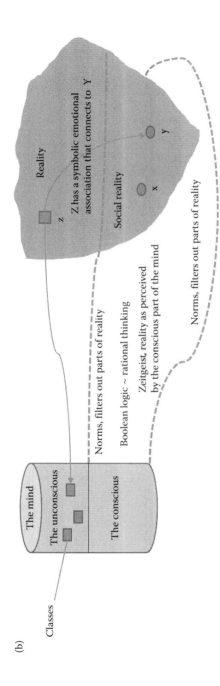

FIGURE 4.2 (*Continued*) (b) If Z has a symbolic emotional association that through a common feature links it to Y, the seemingly irrational decision, from the perspective of the conscious mind, to select Y instead of X occurs.

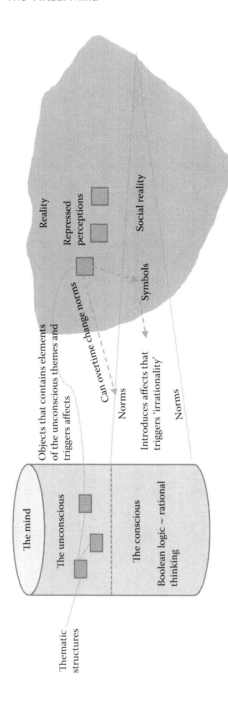

FIGURE 4.3   A depiction of the short-term and long-term interaction between the conscious and unconscious form-ing the holistic mind and its structure of thinking.

- A feed including the full population of perceptions representing the wider set of reality assessed by the unconscious part of the mind.

- A repository consisting of a symbolism taxonomy which distinguish classes, such as taboos falling outside the boundaries of the zeitgeist but that carries an influence capacity when in its covert symbolic form appears within the realms of it.

- A catalogue of classes tracked through time series which can determine trends and help distinguish between reigning themes in the conscious and the unconscious, drawing out potential divergences.

A governing protocol

- A protocol that determines the priority between Boolean logic and the unconscious logic in decision-making queries, drawing on dynamic constants allowing for the establishment for hard coded rules for alterations between the thinking systems.

Chapter 5 will in an object-oriented manner formulate and detail a blueprint of the outlined meta-model. The chapter will also point out previous attempts to create electronic versions of the mind and brain, highlighting where they have fallen short and how this proposal offers a methodology more closely resembling the way humans think.

ENDNOTES

1. Reeder, J. 1988. *Tala/Lyssna. En essä om den specifika skillnaden i Jacques Lacans psykoanalys.* Stockholm/Lund: Symposion Bokförlag.
2. Textor, M. 2010. *Routledge Philosophy GuideBook to Frege on Sense and Reference.* New York: Routledge.

3. Ibid.
4. Lacan, J. 2007. The instance of the letter in the unconscious or reason since Freud. In *Ecrits: The First Complete Edition in English*, translated by B. Fink. New York: W.W. Norton & Company.
5. Ibid.
6. Sutherland, S. 1992. *Irrationality: Why We Don't Think Straight*. New York: Diane Publishing Company.
7. Caplan, B. Rational irrationality: A framework for the neoclassical-behavioral debate. Department of Economics and Center for the Study of Public Choice. UNDATED http://www.gmu.edu/departments/economics/bcaplan/eejformat.doc (retrieved 1 January 2017).
8. McKenzie, CRM. September 2003. Rational models as theories – not standards – of behavior" *Trends in Cognitive Sciences* 7 (9). http://psy2.ucsd.edu/~mckenzie/mckenzie-tics.pdf (retrieved 1 January 2017).
9. Freud, S. 2003. *The Uncanny*. London, England: Penguin Classics, 2003.
10. Tuck, M. August 17, 2010. Gestalt Principles Applied in Design. Six Revisions Website. http://sixrevisions.com/web_design/gestalt-principles-applied-in-design/ (retrieved 1 January 2017).
11. Smith, B. (ed) & Ehrenfels, C. 1988. *Foundations of Gestalt Theory (Philosophia Resources Library*. Munich, West Germany: Philosophia Verlag Gmbh.
12. Tuck, M. August 17, 2010. Gestalt Principles Applied in Design Six Revisions Website. http://sixrevisions.com/web_design/gestalt-principles-applied-in-design/ (retrieved 1 January 2017).
13. Smith, B. (edited by); Ehrenfels, C. 1988. *Foundations of Gestalt Theory (Philosophia Resources Library)*. Munich, West Germany: Philosophia Verlag Gmbh.

BIBLIOGRAPHY

Caplan, B., Rational irrationality: A framework for the neoclassical-behavioral debate. *Department of Economics and Center for the Study of Public Choice*. Undated. http://www.gmu.edu/departments/economics/bcaplan/eejformat.doc (retrieved 1 January 2017).

Freud, S. 2003. *The Uncanny*. London, England: Penguin Classics.

Lacan, J. 2007. The instance of the letter in the unconscious or reason since Freud. In *Ecrits: The First Complete Edition in English*, translated by B. Fink. New York: W.W. Norton & Company, Chapter 20.

McKenzie, CRM. September 2003. Rational models as theories – not standards – of behavior. *Trends in Cognitive Sciences* 7 (9). http://psy2.ucsd.edu/~mckenzie/mckenzie-tics.pdf (retrieved 1 January 2017).

Reeder, J. 1988. *Tala/Lyssna. En essä om den specifika skillnaden i Jacques Lacans psykoanalys*. Stockholm/Lund: Symposion Bokförlag.

Smith, B. (edited by); Ehrenfels, C. 1988. *Foundations of Gestalt Theory (Philosophia Resources Library)*. Munich, West Germany: Philosophia Verlag Gmbh.

Sutherland, S. 1992. *Irrationality: Why We Don't Think Straight*. New York: Diane Publishing Company.

Textor, M. 2010. *Routledge Philosophy GuideBook to Frege on Sense and Reference*. New York: Routledge.

Tuck, M. August 17, 2010. Gestalt Principles Applied in Design Six Revisions Website, http://sixrevisions.com/web_design/gestalt-principles-applied-in-design/ (retrieved 1 January 2017).

An Object-Oriented Architecture Perspective in Designing a Virtual Mind

Artificial intelligence is no match for natural stupidity.

Anonymous

INTRODUCTION

A design solution to a virtual mind requires a bespoke inter-disciplinary approach drawing on influences from psychology, neuroscience, philosophy, linguistics, hardware architecture and programming languages among others. As the goal is to model human thinking from the perspective of the mind rather than the brain, we can leverage from the psychological models that exist and deploy their assumptions as a starting point for the design. That the philosophical notion of a mind and its components does indeed physically exist, neuroscience has provided evidence for; and although one can argue about its various functionalities, parts of the mind can be pinned to processes linked to the brain and the central nerve system. As such, the abstract notion of the mind supersedes the biological processes and therefore serves as a suitable and intrinsically more simplified meta-model of human thinking. While it is of interest to model the brain based on biological material, the challenges to do so are huge and the appropriate technology hardly even exists. Through replicating the functionality of the mind, we avoid much complexity and unknowns that still are associated with the structure of the brain and how the individual brain processes relate and interact.

The virtual mind proposition seeks not to develop machine-generated self-awareness in the spirit of *strong artificial intelligence* but rather to formulate the way humans think under the assumption that it follows rules that can be replicated; thus *weak artificial intelligence*. One has here to distinguish between whether a machine is *really* thinking or is just *acting like* it is thinking, with the modelling efforts aiming for the latter. By including the unconscious part of the mind, it achieves a higher functioning with the capability to interpret data seeking out associations, in accordance with Matte Blanco's logic and gestalt psychology, a capacity that the rationality of the conscious part of the mind lacks. To this comes the realisation that the unconscious absorbs a broader spectra of reality than only the social reality that equals the finite boundaries of rational thinking. By bringing these concepts together into a holistic model governed by a rule-based framework, the ambiguity of human thinking can be explained and forecasted.

However, the traditional focus on artificial intelligence has rested on the assumption of modelling rational thinking and while there has been a recognition of the fact that the unconscious plays a part, it has been assumed that the unconscious recognition of pattern, features such as intuition cannot be captured in formal rules. And this skepticism has prevailed, but by bringing in a multidisciplinary approach and deploying the latest findings from neuroscience, do show that it is possible to assign rules to how the unconscious interprets information and identifies patterns.

This chapter will outline the elements of mathematics and programming languages required for replicating human thinking and to set the context, it will also present the historical efforts to develop artificial intelligence and its objectives. Broadly, these designs have taken the brain, a particular intellectual or cognitive function rather than the concept of the mind as a starting point. From that perspective, some of the attempts have indeed been successful; however, just not in replicating the *holistic way* humans think.

ARTIFICIAL INTELLIGENCE

Computer science has almost since its advent as an academic discipline been concerned with moving beyond the origins of simple mathematical calculations towards ideas of intelligent machines. The term *artificial intelligence* was introduced in

1956. So, what was then artificial intelligence designed to do? It aimed to replicate the ability to learn and to generalise by inferring something which is not part of a stored set of information.[1] As early as 1950, the English mathematician Alan Turing (1912–1954), who is widely considered as the founding father of computer science, asked whether machines can think. The important *Turing test* states that a computer can be considered to be 'thinking' if it can convince a person asking it questions for 5 minutes that it is human in more than 70% of the attempts. Important to note is the distinction that the answers do not have to be correct – they just have to convince the questioner that he is conversing not with a machine but a fellow human. The test is still valid albeit raising many questions and criticisms, especially from a philosophical perspective. No computer has as of yet come close to pass the test, and being able to replicate human behaviour, is just a prerequisite to concluding that a computer can 'think'.[2] However, just appearing to think, is not really thinking. It is, when a computer has been equipped with a suite of mechanistic rules reflecting the actions and interactions of conscious and unconscious logic that is able to replicate and replace human thinking.

Since artificial intelligence had its rooting in computer science, it is often assumed that its applications are designed in hardware and/or software, but it is not strictly necessary. An artificial intelligence system can be modelled in mathematics; a function such as *the least mean squares algorithm* inhibits artificial intelligence by having the ability to fit curves to a set of points and extrapolating new points – a process of learning and generalisation.

One of the first attempts seeking to mimic human behaviour came with the development of a software labelled ELIZA (after Eliza Doolittle) that featured natural language processing capacities, although in a primitive form but which appeared to be capable of reasoning. This application somewhat successfully simulated a conversation with a human, using scripts and pattern recognition matching rather than real intelligence.[3] Pattern matching simply means that it identifies some signal words, such as the 'I need' in the phrase 'I need XYZ' and answers with a pre-set pattern as 'Why do you need XYZ?' Illustrating this with a made-up coding language, it would work in a manner similar to:

IF input_sentence CONTAINS 'I am' THEN

IF First_Time_Written WRITE 'In what way are you' + the_first_word_after_IAM

ELSE WRITE 'Do you want to be' + the_first_word_after_IAM

Another attempt was a programming language called LISP (LISt Processor) created in 1958 and published in 1960, which quickly became the favoured language in trying to build artificial intelligence systems.[4]

Similar logical systems, based on IF-THEN statements, were used extensively throughout the 1960s in trying to achieve artificial intelligence; however, by the early 1970s, the great hopes had turned into disappointment, as the lack of progress had led to disillusion regarding the possibilities to actually being able to replicate human intelligence. And as progress was lacklustre, it led to removal of funding from defense and research agencies; the popularity of artificial intelligence had fallen in vain.[5]

However, alternative routes were tried; already in 1957 the American psychologist Frank Rosenblatt (1928–1971) had worked on a bespoke computer design, *Perceptron*, along the principles of how the human brain works, namely neural networks. Although it showed promising signs and aroused great interest among the general public, it eventually also failed to deliver. In 1969, the American cognitive scientist Marvin Minsky (1927–2016), described the work of Rosenblatt in the book, *Perceptrons: An Introduction to Computational Geometry*. Minsky concluded that early attempts to simulate how a brain works, using artificial neurons, were doomed to fail because they could not emulate the complexities of a biological neuron, such as performing the *exclusive or* (XOR) logic funtion.[6] During the 1970s, Marvin Minsky would continue to work on a framework in which the mind is seen as a number of individual processes, or *agents*, culminating in his influential bestseller, *The Society of Mind*. In the book, he provides a philosophical perspective, but much of his writing is also applicable for programming as it elaborates on the idea that a virtual mind could be developed through combining a large number of independent small processes into a holistic system.[7]

EXPERT SYSTEMS

By the 1980s, artificial intelligence was yet again coming in vogue, in particular with the development of expert systems, pioneered by the American computer scientist Edward Feigenbaum, who proclaimed that the world was moving from data processing to knowledge processing. An expert system is designed through the use of logical rules and as such apply reasoning to input. This logic engine should also *theoretically* be able to deduce new knowledge from existing facts and rules and thus act more like a human expert being able to think 'new things'. Expert systems are trained so that questions and responses are taught in a hierarchical data structure. The logic focuses on learning and one has to train the expert system on what one thinks are good answers, which is done through an iterative process where the expert system is exposed to a large number of data, highlighting the correct answers with the exact methodology depending on particular system and algorithm. The key thing to note is what it considers the correct answer is based on the data *deliberately* chosen. After it has been trained through the examples, one can test the expert system with a new suite of data to ascertain whether it recognises the correct answer or not. Users can then interrogate the application which will guide the dialogue through previously stored questions and response patterns, directing the user by way of what subject matter experts have programmed into it. When the expert system finds a solution, the iterative dialogue ends but if it cannot find any solution, it will proceed to ask the expert for the resolution. In order to obtain acceptable results, it is vital that the expert system is trained with large amounts of data, which make the process of learning to identify the data and patterns highly resource intensive. And of course, if we by mistake have fed the expert system with wrong answers, it is not able to self-adjust but the logic engine will deduct erroneously. The expert system does not really grasp the concept of what a correct answer is; it is simply just an algorithm reviewing large datasets, identifying patterns and based on these determining what is logically right or wrong. Attempting to replicate the human mind this way, through assembling thousands, if not millions of scenarios for large-scale testing, will still not allow us to understand how the mind works as it is not developed for general problem solving and decision making, and sooner or later some unanticipated scenario would occur that would make the model fail. A second major flaw is that their learning process is dependent on what was taught and the selected reference data, and a poor quality of input will only yield poor output. As an example, an expert system being trained on the jargon from biochemical journals and drawing conclusions thereof would struggle to respond in the conversation with a child, for instance.[8,9]

In the early 1990s, these specialised expert systems were broadly regarded to have failed as they proved too difficult to program and was not able to deduce new knowledge to the extent of early expectations. Although ELIZA failed to make an impact, it had laid the foundation for an important new field; *natural language processing*, or NLP. NLP describes how to derive meaning from human communication and the understanding of natural language which often deviates from logic structures by containing idiosyncrasies and ambiguities. The research area has progressed from a slow start where its first focus was to build machine translation from logical program code based on grammatical theories presented by the American linguist Noam Chomsky, which however yielded few, if any viable results. But, as more powerful machine learning tools have evolved, NLP has made steady strides, creating tools like *Google Translate* that provide translations getting close to being understandable but still far from accurate language.

NEURAL NETWORKS

Neural networks have become one of the most important techniques in the facilitation of learning machines to think. Neural networks, or rather artificial neural networks, are attempts to create an electronic brain by trying to replicate neurons, hence its name.

It was introduced and tested as early as 1943 by the American neurophysiologist Warren S. McCollough (1898–1969) and the American logician Walter Pitts (1923–1969) with a view to mirror the biological neural networks of the brain; obviously as a highly simplified simulation. The idea was based on activated neurons, as these are connected to another layer of neurons and so on, eventually producing an output. Essentially, the input units will react to the given input and pass it on to hidden processing units and if this hidden unit is stimulated by enough activated input units, it will itself activate and pass on a signal that is brought forward in the neural network and when reaching the output units, a result is returned.[10]

In essence, neural networks have an inductive and bottom-up problem-solving approach. The previously mentioned Perceptron was a neural network modelled from a biological equivalent, where the neuron is composed of inbound connectors (or 'input synapses'), an internal aggregation, an activation function and outbound connections (or 'synaptic terminals'). The aggregation function will apply a configurable synaptic weight to the input signals which is calibrated during training. A single or multilayered network is trained using a back propagation algorithm to adjust the synaptic weights by being presented with input and output patterns. While the Perceptron, like that of other neural networks architectures, are useful designs, they are in the context of human thinking only the infrastructure, as they do not offer any higher level of cognitive functionality. For instance, an image recognition neural network can learn to correctly identify breast cancer in x-rays or heart conditions from ECG prints, but will not elaborate on the causes of these, what medical impact they have for the patient or what remedy can be used. Through training and pruning, the inductive neural network gradually learns how inputs relate to outputs and this knowledge is stored in those synaptic weights distributed through the network. When presented with a new sample of data, which has not been previously learnt, the neural network will respond with the closest matching output it can approximate based on the training samples. It is therefore not possible to reverse engineer the knowledge stored in the neural network similar to that of a mathematical formula, database tables or software code. Expert system on the other hand is deterministic in that algorithms or program code are written with advanced knowledge of the domain and applicable rules, which are designed for a particular problem only. For the expert system, knowledge will be stored in a structured format and retrievable through a sequence of interrogations to arrive at a result which one would expect that the expert in the domain should achieve.[11]

However, it took until the 1950s for neural networks to be tested in computers. Some early successes appeared in systems to reduce echoes in telephone lines by predicting the correct sound, but neural networks failed to make greater impact in industrial applications and like many other techniques used in artificial intelligence, it eventually lost popularity. But with more recent access to *big data*, neural networks have yet again become a popular technique, as any neural system now can be flushed through with huge volumes of data, and the hidden processing units be trained to detect patterns and their connections in these large batches of data, otherwise difficult to identify. Out of this, has the concept of *deep learning* emerged which helps to identify patterns in seemingly large amounts of unrelated data points. Deep learning is conducted through a collation of algorithms, in effect taking a similar approach as the neural network's hidden units that connect to each other and activate signals for usage in other (output) units. But for deep learning applications, each neural network or function becomes a unit in itself, allowing for a more in-depth analysis as different neural networks cooperate.[12]

The neural networks and deep learning networks can be said to originate from a desire to physically build an electronic brain that works in a similar way as our biological brain. Examples include the *Human Brain Project*, a (controversial) 1 billion Euro bet to build a replica of a brain inside a computer,[13] or a smaller project to simulate a roundworm's brain of about 500 neurons,[14] or the *Blue Brain Project* to reverse engineer a rat's brain.[15] It is however worth pointing out that the structure of neural networks and the neuroscientific understanding of the brain have moved quite far apart over the years. As our understanding of the brain has increased, at the same time the neural networks models have been made more efficient for their tasks in processing data, which absurdly has come to mean that they have been deviating from the updated knowledge of the functions of the brain.

So, in the attempts to design a machine-generated replica of human thinking, one can distinguish between two main paths; mimic a brain with its neurons or try to simulate the more elusive concept of the human mind. The approach to design could not be more distinctively different with either the path of defining a suite of rules that mechanically can reflect human thinking or creating all the processes that the brain consists of. Electronically replicating a brain becomes an enormous undertaking, the approach to replicate the mind is considerably less onerous; however, it has so far been unsurmountable as no one to date have been able to pin down that suite of rules. And just by electronically reproducing all the neurons in a brain does not guarantee it will generate a model that behaves similar to that of a human brain; the workings of the brain are still far too unknown and complex for a computer with a mimicking structure to be able to reflect it. However, the early example of ELIZA, trying to resemble natural language shows that 'simple trickery' almost can convince people of a machine being able to think.

The discussion often comes back to the question of what we consider to be a mind, and what we want to use it for. For the purpose of this book, the proposition is to build an application that can simulate how a human mind works in a collective proxy setting, rather than a technical reproduction of any biological properties.

RECENT DEVELOPMENTS

A renewed popular interest and fascination has appeared for machine-generated thinking, artificial intelligence and the philosophy surrounding this.

No doubt this is driven by the amount of online data available and how computers and mobile devices have integrated with everyday life. One controversial person responsible for the reawakening of this discussion is the American computer scientist Raymond Kurzweil who has published bestselling books since the 1990s but recently has been associated with the concept of *technological singularity*, which draws to forefront the implications of what will happen when 'real' artificial intelligence becomes a reality.[16]

One of the key issues is that these artificial intelligences are likely to be able to reprogram themselves, and the acceleration of it will eventually be far too complex and result in an intelligence explosion and a digital 'superintelligence' – one that we might not even be able to understand.

This scenario poses philosophical problems; it could eventually threaten mankind itself as argued by the Swedish philosopher Nick Bostrom in the book, *Superintelligence: Paths, Dangers, Strategies*.[17]

As of now these discussions are mainly in the realms of philosophy rather than everyday life – but the fact that they have started to enter the broader debate implies that we are living in a cycle where computers not only promises to be able to understand and simulate our minds, but that we are now contemplating what could happen beyond that.

MATHEMATICS AND TRADITIONAL LOGIC

Artificial intelligence is developed using a high degree of mathematical elements paired with system theory. With the objective of being able to demonstrate repeatable results given the same conditions and inputs, artificial intelligence has progressed to apply a strictly structured and rational approach and to this the modelling consists of mathematical rules, operators, functions and logic, including:

- The Commutative Law: $a + b = b + a$

- The Associative Law: $(a + b) + c = a + (b + c)$

- The Distributive Law: $a \times (b + c) = (a \times b) + (a \times c)$

- The Transitive Law: $a > b, b > c => a > c$

- Boolean logic resulting in *True* or *False* depending on the operators NOT, AND, OR, NOT, NAND (Not AND), NOR (Not OR), XOR (Exclusive OR)

- Assignment and comparison operators and functions e.g.: $=, ==, <, >, <>$

- Set Theory IN, NOT IN, JOIN, INNER JOIN, OUTER JOIN

- Functional theory and geometry, e.g., SQRT(X), X^3, EXP(X), SIN(x), COS(x), TAN(X)

- Statistical functions, e.g., MEAN(X), MIN(X), MAX(X), VARIANS(X)

- Probability theory, e.g., PROB (X > Y), RANDOM(X), probability density functions, parallel and serial probabilities

- Linear algebra e.g. vector and matrix calculations, $A = M \times B$

- Single and multivariate analysis e.g. derivation, integration, differential equations, Taylor approximation.

The list of mathematical theories and methods applied is extensive but these are utilised with the purpose of providing fundamental, repeatable and rational expressions about the objects or relationships implicated.

PROGRAMMING LANGUAGES

Programming languages have had a rapid development since the inception of the semiconductor, transistor and the digital computer. Originally, programing languages were bespoke instructions designed specifically for a particular microcomputer architecture, which were built around a particular CPU and chip set type. Popular architectures, intermediary 'monitor' programmes and operating systems evolved, which abstracted the underlying hardware away from the programmer so that more focus on functionality could be made.

Many programming concepts and languages developed and are still evolving but there are some fundamental building blocks and a few milestone designs have become standard:

- *Data structures and types*: Integers, floating points, Boolean and texts could be modelled distinctively with physical memory size and relevant functionality associated.

- *Permanent storage of data*: Perhaps not a paradigm in itself but a very significant development of storing working data between sessions in files and databases.

- *Modular*: Often reoccurring instructions should be collected in a procedure or function so that it can be repeatedly called on as needed.[18]

Interpreters and compilers progressed such that programs written in a structured human writable language could be converted to an operating system with readable instructions set, which are based on procedural language rules[19]:

- *Sequence*: A software program runs in a sequential process usually executing commands top to bottom, left to right.

- *Selection*: The direction of processing can be changed in run-time dependent on specific criteria.

- *Iteration*: A process can continue for a fixed or indefinite number of times.

With the introduction of *object-oriented programming*, coding took a revolutionary step forward. This paradigm shift allowed for more complex and scalable software to be designed. Object-oriented programming is based on three fundamental concepts[20,21]:

- *Inheritance*: Similar classes allowed for functional reuse by designing a general 'super class' and more specialised subclasses.

- *Encapsulation*: Objects can have internal functionality only accessible to themselves or public classes accessible by any other class. This allows a software module to have small, specific footprints exposed even though there may be a complex algorithm contained.

- *Polymorphism*: Depending on what data are processed by a class, the functional behaviour can vary. This allows the design of a module's signature to be simpler and more generic.

Object-oriented technology is now a standard feature of any software development. All mature languages generally have support for object-oriented programming, input/output, mathematical functions, graphical user interface and database development. While they differ in programming syntax, platform and enterprise application support, their functionality are largely similar.

Early artificial intelligence practitioners appeared to believe that the programming language chosen was key to unlocking higher cognitive capability. Perhaps this was because computer scientists got so caught up in designing a highly functional and perfect programming language that they began to relate it with the capabilities of a human brain or mind. LISP has often been the language of preference due to the data structures and occurring recursive designs; however, any modern programming language, in particular object-oriented language, is easily capable of handling these features.

DESIGNING A VIRTUAL MIND

All the discussed functions and methodologies have been, to various degree, successful at what they were aimed to do, but none of these was purposely intended to create a model of human thinking capable of integrating conscious and unconscious logics as well as subjectively disregard or reinterpret parts of reality. But we cannot use most of the above functions or traditional logic to model unconscious logic due to their strict mathematical rooting.

In the design of a virtual mind, one must distinguish the mind from intelligence, as intelligence itself is not the only capability the mind holds, unconscious and emotional behaviour are the most obvious features that need to be incorporated in any model attempting to describe and predict the activities of human decision making. To reflect the mind, methodologies that are underpinned by logic inference will alone not suffice, as we know that the conscious part operates rationally on a truncated view of reality and that the unconscious relaxes these requirements, absorbs broader parts of reality and allows for a seemingly more obscure albeit still structured mental framework. To note is that we are not trying to build a virtual personality, awareness or spirit, hence the Turing test is therefore not relevant as the design is not an application which will seek to pass as a human being but rather aims to replicate and forecast human thinking.

To create a virtual mind, we have the optionality to design it as digital hardware, with a future optionality to migrate it to hardware, or in software. The following section will provide generic descriptions of both options, and discuss pros and cons of each.

Hardware

Analogue hardware is ideal for systems that natively operate in a continuous range of power levels such as in a physical sensory and actuator environment. Applications suitable for analogue hardware are typically robotics, weather stations and audio amplifiers. However, over time these applications have become more digitalised because of the accessibility of digital components, microcomputers and embedded computers that allow for standardised input/output interfaces, analogue to digital converters and logic written in software rather than hard wired in the electronic design. The benefits with analogue hardware are as follows:

- *High speed*: Electrical signals travel at near light-speed and there is nothing impeding the communication between components.

- *Resiliency*: Analogue hardware can be un-influenced by disturbing electrical and magnetic fields, provided proper shielding. In radiation-intensive environments such as outer space, there are high-energy cosmic radiation that will cause disastrous bit-flips in digital hardware which do not occur in analogue circuits.

- *Small size*: Circuits can be designed very small, limited only by the size of the sensors and actuators. Integrated circuits may contain many functions in a small electronic component.

- *Low power*: Circuits do not need to consume electricity at all unless they are used actively.

There are, however, a few drawbacks with analogue hardware:

- *Functionality*: There is limited functionality that can be implemented practically. A complex function may require a lot of components and hard-wired electronic design to craft.

- *Complexity*: Designing analogue circuitry requires expert electronics knowledge for prototyping and for mass production.

- *Time consuming*: It takes a relatively long time to design, build and test circuits compared to digital or software options.

- *Cost*: The upfront and ongoing cost of building analogue circuitry is high compared to software.

Digital hardware is realised by logical circuits, microcomputers and embedded computers. Digital signals are either logical 1 or 0 and these signals are in nearly all cases communicated using a synchronised clock frequency so that operations are carried out in a controlled and consistent manner. The signals are communicated between components using a serial or parallel interface. Numbers are represented using base two-binary format. Benefits of digital hardware include:

- *Capability*: Complex applications are more readily designed in digital hardware than analogue hardware. Circuitry in field programmable gate arrays (FPGA) or VHSIC (Very High Speed Integrated Circuit) hardware description language (VHDL) can be realised for significantly complex applications ranging from signal processing through to artificial neural networks.

A drawback similar to analogue circuity is

- *Complexity*: Designing digital circuitry requires expert knowledge of electronics for prototyping and mass production.

Software

Software allows for functionality to be written in a higher abstract language more similar to human language than what either analogue or digital hardware can provide. Developing in software has the benefit that functionality can rapidly be written, tested and released to production. The programming languages and software components available today support effecting human interface, logical operations, mathematical and statistical operations and input/output communication. Application software runs on many layers of components and infrastructure programmes which allows for highly complex functionality. While the speed of software cannot compare with the faster hardware circuits, the usage of multi-threading, distributed grid computing or other parallel computing, distributed data storage and processing schemes allow for much greater 'on average' speeds which holds the ability to exceed hardware speed.

THE FUNCTIONAL MODEL

The functionalities of the virtual mind are, at the core, to demonstrate conscious and unconscious logic and a governance protocol, which interpret the states of the conscious and the unconscious and determines what logic of thinking should apply depending on scenario and input. Figure 5.1 shows the functional model of the virtual mind.

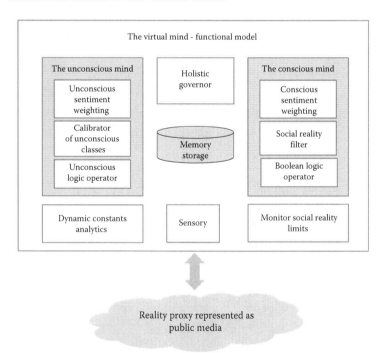

FIGURE 5.1　Functional model.

Sensory [The Reality Perception]

Rather than using sensors similar to the five human senses, the virtual mind will approximate reality by searching public media. The applications are therefore limited to non-physical sensory absorptions. An ongoing feed of reality perceptions is needed by drawing from broad-based enough media sources to represent a proxy of the collective mindset. The sensory function is an extract, translation and load (ETL) device which provides input to the conscious and unconscious thought capabilities. A limited set of perceptions will be processed by the conscious logic device as truncated by the social reality filter; however, the full population of perceptions will be absorbed by the unconscious device, aligned with what has been confirmed by neuroscience. Data from the public media are interpreted and the output is stored in the memory storage and passed onto the conscious and unconscious mind, respectively, for more refined processing.

The Unconscious Mind

Unconscious logic operator [symmetric logic]. The calculation device for the unconscious logic is based on the principles by Matte Blanco's symmetrical association and its mechanisms. It serves the fundamental capability to interpret and assign symmetric formations using a suite of class structures.

The unconscious mind operator will search the public media contents for sentences which contain symbolic language. The symbolic contents will be allocated and clustered and mapped in preset classes that are hard coded into the functions. These are then counted for occurrences and ranked in terms of frequency by time unit to establish trends. A more frequent usage of symbol words will have a higher stimulus intensity.

Unconscious sentiment weighting. The unconscious incorporates prevalent emotions with the symbolic perceptions. The emotion associated with a particular symbol is determined from the language used in the matching text strings over a period of time and the intensity of the emotions is measured, through its frequency in a simple count. The weighting is the count of emotional words e.g. *'better'*, *'worse'* that occur in text strings which contain figurative language. The weighting is later applied to the unconscious symbol words to augment or suppress the overall symbolic responses. This is calculated statistically over the period using all text sources as inputs. Through the frequency analysis, the symbolic language and its classes and attached sentiments hold the propensity to reflect human ambivalent thinking. In its simplest versions, the sentiment rating will be binary associated with either a positive or negative value. The unconscious response will be linked with a positive or negative value so that it can mathematically complement the conscious response. For instance, the occurrence of the word 'better' is associated with positive '+1' and the occurrence of the sentiment word 'fear' with negative '−1'. This will allow the holistic governor to calculate the aggregated response from both the conscious and unconscious and through the combined count of these determine the strength of the response, which will dictate the outfall of the decision making.

An example scenario is the idiom *'The grass is always greener...'*, appearing in one of the text sources as picked up by the ETL device from the proxy reality. The symbol words here are *'grass'* and *'green'*. But to gauge these symbols' impact on the mind, one needs to determine its sentiments, which is done through filtering out emotion words occurring in the same text source as *'The grass is always greener...'* and attach them to these symbol words. Assuming that the emotion word *'better'* appears, the unconscious response to this individual text source and its score is calculated as in the table below:

Symbol	Count	Descriptor	Sentiment
Grass	1	Better	Positive, +1
Green	1	Better	Positive, +1

The symbols are then clustered to a predefined suite of structures together with related symbol words and then ranked for frequency for the particular time period, which is then added to previous time periods to form a time series. Paired with these are the associated sentiments.

In aggregation for that specific time unit, let us say the following symbols are measured and sentiment weighted:

Symbol	Count	Sentiment Weighting
Grass	15	+5 positive
Green	3	+1 positive and −3 negative

Unconscious response (Grass) $15 \times (+5) = 75$

Unconscious response (Green) $3 \times (-2) = -6$

Calibrator of unconscious classes. To arrange the unconscious perceptions, these are structured around classes. The classes are finite in numbers and will in accordance to Matte Blanco's rules aspire to unity, so that they will seek to cluster around these. To identify their contents, all symbolic languages are collated and analysed for common denominators and these make up the classes to which future symbols are mapped. Each class is tracked through time series, where all corresponding occurrences are monitored for increasing trends, that might form the contents of an emerging zeitgeist.

The Conscious Mind

Conscious logic operator [asymmetric (boolean) logic]. The calculation device for the conscious part of the mind is the typical Boolean logic, or similar, to represent the asymmetric structure of rational thinking. This function performs a logical process of computing output results from input information belonging to the social reality domain, with the output being the conscious response. The social reality filter will reduce the amount of information forwarded to the conscious logic operator depending on the prevailing class trends which are applied as boundaries. The logic operator calculates the conscious response by aggregating all the rational stimulations generated from the input information and this will be dependent on the specific query the virtual mind is tasked with.

Conscious sentiment weighting. The conscious responses need to be further amplified or reduced by sentiments that have accumulated over a period of time. The weighting is the count of sentiment words occurring in the text sources e.g. 'better', 'worse' where symbolic language does not occur simultaneously. This is calculated statistically over the period using all applicable text sources as inputs.

Social reality filter. The conscious part of the mind needs a mechanism to reduce the large amounts of perceptions based on reigning trends. The social reality filter is able to repress perceptions deemed not suitable due to societal norms and moral standards. This filter is applied on the perception feed to distinguish the social reality from the physical reality, constructed to highlight the classes allowed for by the zeitgeist. We can measure this exactly through a count of occurrences over time and focus on the popular classes as broadly defining the boundaries out of a multitude of various differing views.

In other words, what is mainstream becomes social reality. Items falling outside of these will be absorbed and registered by the unconscious and might still play a part in the decision-making process. The filter will reduce the amount of information forwarded to the Conscious Logic Operator function based on the prevailing social reality classes. If contents are presented which are not part of a prevailing social reality then they are disregarded immediately and not processed further; however, they are always processed in unconscious theme calibrator function.

Holistic Governor [Integration Logic]

This is the protocol that determines the priority between Boolean logic and the unconscious logic and consolidates them in decision-making queries, drawing on configured dynamic constants allowing for the establishment of hard-coded rules for alterations between the thinking systems. The conscious responses are complemented with the wider perceptions of the unconscious absorptions, evaluated and aggregated by the holistic governor, which may lead to seemingly irrational responses. Mostly we will expect a rational response, however, in some cases, an affect perceived irrational response will occur where there is an overwhelming unconscious element which contradicts the rational, and the ambiguous irrational response will override. The levels of such overrides are determined through back testing of historical data. Care in design must be taken so that the Boolean logic can be re-enforced or contradicted by the unconscious logic, if not the system may start to demonstrate symptoms of psychical ailments such as becoming schizophrenic.

The aggregation associates the sentiments linked with the social reality classes in determining a response, so that a positive emotional sentiment will precede over a negative sentiment, which is the expected rational response in the choice between two options. The frequency of occurring sentiments will be decisive in the decision-making process, such that a higher number of positive sentiment responses for a class will rank higher than that of the one with less positive sentiment responses.

In cases where the text mass analysed for a particular time period contains symbolic language with classes that contravenes the boundaries of social reality, these are incorporated in the holistic response to capture the ambiguity of human thought processes. The holistic governor integration function will calculate the final response by aggregating all the unconscious and conscious responses using a weighted aggregation function. The result will be stored in memory alongside other data.

Holistic response = Aggregated conscious response and unconscious response

$$HR = \sum (CR + UR)$$

Holistic response = (Conscious response + unconscious response)

This can be illustrated with the example of deciding between a red and a blue car. Assuming within the text mass and designated time period the combination blue car and positive emotional sentiments have occurred 50 times, and with a negative emotional sentiment 10 times, and the combination red car and positive emotional sentiments have occurred 80 times and with a negative emotional sentiment 5 times.

Conscious response = Blue car (+50 −10) versus red car (+80 −5), e.g., blue car 40 versus red car 75.

Thus, the rational choice will be the selection of the red car.

But a holistic response would need to include the unconscious sentiment as well to better reflect the way human thinks, so that:

Holistic response = (Conscious response + unconscious response)

Assuming that red, in addition to its meaning as a colour, also holds a symbolic meaning that relates to a class that has been repressed into the unconscious due to it being regarded as a breach against reigning norms, whether of a sexual, political or other inclination, is a taboo of sorts. Blue on the other hand in its symbolic meaning remains indifferent in the holistic context. Hence, the appearance of the word red in its symbolic form appearing in the same text mass and the same time period must be considered. Important to note is that all topics must be included, not only articles covering cars, but also topics seemingly completely unrelated, such as fiction, sport and culture. Assume that the red in its symbolic meaning occurs 60 times:

Holistic response = (Conscious response [blue car 40 versus red car 75] + unconscious response [red -60])

Holistic response = Blue car 40 versus red car 15

When considering information outside the constraints of social reality, the choice becomes to prefer the blue car.

Dynamic Constants Analytics

Through measurement of historical data, constants, also of a dynamic nature, can be established. These would relate magnitudes in terms of intensities and frequencies for the overriding of responses including unconscious contents as well as changes in classes that define the zeitgeist. An ability to view, search and calculate statistics of these repository of various constants is needed, as they will allow for projections.

Monitor Social Reality Limits

To understand and calibrate changes in the social reality, a mechanism to enhance or reduce perceptions based on prevailing trends is part of the holistic virtual mind. This is achieved through the ongoing monitoring of popular classes in public media including their associated sentiments that over time tends to shift, as a certain class exhausts in popularity and eventually falls out of fashion, to be replaced by something else. And backtesting allows for calibrating and establishing dynamic constants when new social reality classes emerge and old ones deteriorate and become obsolete. By measuring the prevailing trend and acknowledging that these will change over time, albeit that the intensity and duration might differ, thereby the 'dynamic' attributions to constants as they might need adjustments. The trends are calculated in the analysis function using all media resources in a period, and applied for each applicable text source to determine changes. These dynamic constants can take two major forms: in duration and in relative frequency, the strength of the class impact on social reality. A repository consisting of a symbolic taxonomy which distinguish classes will help to highlight the ones falling within the boundaries of the social reality and the ones not, such as taboos, but with the propensity of carrying an affect capacity when in its covert symbolic form appears within the realms of the social reality. Dramatic and rapid changes in social reality classes can occur when a build-up of conflicting classes exists between the conscious and unconscious. When this happens, the classes will be recalibrated so that the virtual mind better matches reality. The lead up to recalibration is indicated by an increase in unconscious responses.

Memory Storage

The memory storage is a fundamental feature needed to persist holistic, unconscious and conscious responses as well as holding the repository of dynamic constants. The class structures are also stored here.

DATA MODEL

Figure 5.2 presents the data model needed to store input, the intermediate and response data. The mappings and configurations of the data are needed to train responses based on existing dynamic constants, facilitate responses due to unconscious input as well as general application configuration.

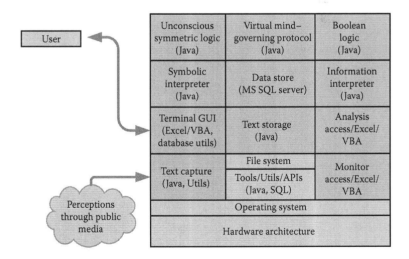

FIGURE 5.2 Data model.

Architecture of the Data Model

The virtual mind capability and functional requirements can as previously discussed be implemented on many alternative platforms. This example will describe implementation in software as the sensory inputs in public media data are in digital format and the outputs are data points of various measures which greatly facilitates software as the preferred option. The software implementation also has the advantage that the model can be flexibly and quickly enhanced, as software runs on operating systems and digital hardware, applications can easily be designed to exhibit non-linear, random and irrational functions that are the observed properties of the mind. The choice of programming language is an object-oriented language as it provides scalable and flexible features.

The core processing logic is almost entirely made up of lookups from static data tables. The technical stack uses relatively standard easily available components, and PC equipment as the virtual mind will operate in a digital computing environment with interfaces to filesystem, database, Internet and console. Thus, the design solution includes the typical business application with a data model and software architecture approach by creating a capability capacity including the static and the dynamic functions first. Following that, the technical architecture and software stack meeting the functional requirements are the next development steps. The next steps include drafting the program code and data configuration, which will be concluded by testing the software components and commencing the back testing to gradually calibrate and improve its performance through supervised learning.

The public media will be captured using utilities that download text resources and is sourced from the Internet using Access/

VBA functions and stored in the SQL Server for further symbolism and information interpretation. These media sources, which come in HTML format, are configurable in the database and are adapted to be filtered and enriched by the information- and symbol-interpreter functions.

The core processing functions, that is, the conscious logic, the unconscious logic and holistic governor will be designed in Java, suitable as a highly capable object-oriented programming language with a large set of APIs (application program interfaces) for many types of applications. An SQL Server relational database management system (RDBMS) supports data storage, mapping and configuration. The analysis component is in its first version implemented by VBA Excel and Access using the MS SQL Server as a data repository. These software tools will run on a regular Windows PC.

The Logic Functions

The symmetric logic operating in the unconscious part of the mind is a function of associations as outlined by the principle established by Freud and Matte Blanco and described in previous chapters. It evaluates whether one symbol is similar to another based on a repository of symbolic classes. These classes are finite in nature as they represent the typical situations in life and are represented in a dictionary *cum* lexicon that serves as the mapping structure to which figurative language is categorised from the text sources. This dictionary will obviously have to be updated over time in order to add newly minted synonyms to the core symbols. The classes will be tracked over time through assigning the time-stamped occurrence of figurative language as a time series highlighting a measure of intensity and its trends. So rather than single symbol words, it is the classes that are tracked, to avoid the idiosyncrasy through a symbol word randomly appearing in a non-symbolic manner and distorting trends.

The association is conducted through a matching algorithm that compares specific matching fields for each class which in its traditional function applies operators such as =, < >, including levels of tolerances. But as already argued, the unconscious does not use these operators in symmetric processing. And another concern is defining matching fields, such as on the symbol class 'war' with 'red' in its symbolic meaning as blood rather than love *vis-à-vis* than its actual meaning as colour, and thereby allowing for accurate matching. Strict matching algorithms will in this case provide a diluted value; instead a classification, or clustering, method will better meet the requirements. Each symbol word will be assigned to a more abstract meaning, in addition to its dictionary meaning. This form the generalisation – and aggregation – process, which in addition to its classes serves as the atom level, which allows for escalating up to unifying class. To this effect, mapping tables provide the template, so that when a symbol is encountered, a count to the corresponding class is added. Where a symbol, in its *Fregian* fashion can hold multiple meanings, correlation tests with related class words are done, to statistically highlight the specific meaning. Understanding the impact of the symbol classes is through the intensity measure, where each class can be ranked and its importance graded.

But also, the classes' emotional impact needs to be understood, in its most simplest form, through a positive or negative mood. These emotions clustered in the text mass with the symbolic language functions as additional weighting to the symbolic classes, so that the symbols will be associated symmetrically to an emotion and contrasted to the views and opinions of the collective conscious mind. The particular emotion that the symbol class has stimulated is therefore relevant for further processing in the virtual mind's decision-making process. Again, this is ranked by intensity through counts and the plethora of emotions are sourced from a lexicon in a mapping table. A symbol class will map N:1 with the conscious response where occurring, so that it can be included and provide a holistic response. Hence, every emotion types will have several symbols associated with it. Through ongoing testing of the linkage between symbolic classes and emotion types, the function demonstrates intelligence, as artificial intelligence defines it, by incorporating these learning and generalisation capabilities, ultimately forming dynamic constants. This learning function is demonstrated by the ability to configure and re-calibrate these areas as new data are included, such as the raw text mass to the symbolic language mapping and the related emotions mapping.

The generalisation capacity is demonstrated through the association of symbols into holistic response, where the unconscious is matched with the conscious and aggregated. The elevation process of classes is conducted through pairing classes that can be grouped to the same reigning emotion types, such that if the contemporary symbol classes belong to the same emotions type, then the symbols are regarded the same. Therefore, there is no need for attribute-by-attribute comparison as the generalisation, according to Matte Blanco, matches seemingly unrelated symbol classes together through the spectra of emotions. The seemingly unrelated classes are held together by the specific emotions. This constitutes the abstraction process, simple and not always rational – as would be expected by the unconscious – and evidenced through the unconscious aspiration towards symmetry to stereotypes or archetypes depending on perspective. It is assumed that the sensory inputs, proxied through the public media, and the responses they generate eventually will have an impact on reality, such that a divergence in conscious and unconscious classes will be reflected in, over time, a changing social reality.

Later versions may include options of incorporating or implementing the entire virtual mind system on analogue or digital hardware as it will allow for a broader range of applications embedded in non-digital environments where analogue electronic circuits for sensory and actuary controls can be integrated. However, it would require a need for significant low-level circuitry for integration and communication between parts but following this there are many electronic sensors that could be used to augment traditional applications.

FLOW MODEL

A flow model of the virtual mind is shown in Figure 5.3.

1. Reality is that represented by public media. Obviously, a crude proxy to 'physical' reality, but this domain serves as a simplification of physical medium and captures the data that represent the collective holistic mind containing themes of social reality, the figurative language representing the unconscious undercurrents and the various emotional sentiments. The flow of data is housed in a data storage.

2. The ongoing flow of text mass is directed to the unconscious part of the mind, where it is sorted and categorised through a filter of classes and associated emotional sentiments are mapped.

3. Symbol class association evaluator. With symbol classes as measurement units being used to measure the intensity of activities in the unconscious. These are ranked through trending in assessing popularity. In any given time, several symbol classes are expected to exist, representing the ambiguity of the human mind and mixed responses are possible, even likely. These are linked through shared emotional types and can at onset be seemingly irrational. But generally, the expectation is the weighting towards a single particular response. Previous mental state will affect the current evaluation to achieve a 'lag' in change.

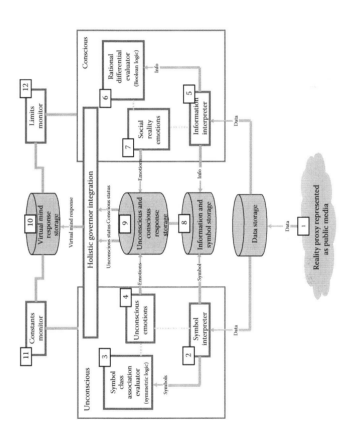

FIGURE 5.3   Functional flow model.

It will compare with previous holistic responses for re-enforcement or offsets capturing a momentum aspect. The set of unconscious response and their intensities are forwarded to the holistic governor.

4. Emotions associated and mapped to unconscious classes are tracked to be contrasted in the holistic governor.

5. The ongoing flow of text mass is also directed to the conscious part of the mind, where it is sorted and categorised through a filter of social reality classes and associated emotional sentiments are mapped.

6. Rational differential evaluator. Social reality classes are compared and ranked through Boolean logic operators, in such a way that the selection between two social reality classes are ranked in magnitude. The Rational Differential Evaluator will compare with previous holistic responses for re-enforcement or offsets capturing a momentum aspect. A state defining the social reality specific to the domain is the result. These states are forwarded to the holistic governor.

7. Emotions associated and mapped to social reality classes are tracked to be reviewed and ranked in the holistic governor.

8. The unconscious and social reality classes are stored in a repository of patterns of classes in terms of magnitude and duration to facilitate the establishment of dynamic constants.

9. Similarly, for emotions associated with unconscious classes and social reality classes, these are stored, as future reference to develop a training mechanism so that constants can be established.

10. The holistic governor receives the unconscious and the conscious interpretations of the sensory data. The governor integrates both states into a holistic state of mind. The

unconscious and the conscious states could re-enforce each other, cancel out or contradict each other. The resulting response will be dependent on the domain and the respective intensities for each states.

11–12. Function to analyse constants for unconscious classes and social reality limits are tested and calibrated through the feeds of data and its classifications and is used to determine the outcome of the responses. This is used for backtesting of models which will provide calibration and configuration information to the virtual mind's processes.

ENDNOTES

1. McCarthy, J; Minsky, ML; Rochester, N; Shannon, CE. August 31, 1955 A proposal for the Dartmouth Summer Research Project on artificial intelligence. The 1956 Dartmouth Summer Research Project on Artificial Intelligence, Bell Telephone Laboratories. http://www-formal.stanford.edu/jmc/history/dartmouth/dartmouth.html (retrieved 1 January 2017).

2. Turing, AM. October 1950. Computing machinery and intelligence. *Mind* 59 (236): 433-460. http://www.jstor.org/stable/2251299 (retrieved 1 January 2017).

3. Weizenbaum, J. January 1966. ELIZA – A computer program for the study of natural language communication between man and machine. *Magazine Communications of the ACM* 9 (1): pp. 36–45. http://dl.acm.org/citation.cfm?id=365168 (retrieved 1 January 2017).

4. McCarthy, J. April 1960. Recursive functions of symbolic expressions and their computation by machine, Part I. Massachusetts Institute of Technology, Cambridge, MA. http://www-formal.stanford.edu/jmc/recursive.pdf (retrieved 1 January 2017).

5. Lighthill, J. 1973. Artificial intelligence: A general survey. In Artificial Intelligence: A Paper Symposium, Science Research Council. http://www.chilton-computing.org.uk/inf/literature/reports/lighthill_report/contents.htm (retrieved 1 January 2017).

6. Minsky, M; Papert, SA. 1969. *Perceptrons – An Introduction to Computational Geometry.* Cambridge, MA: MIT Press.

7. Minsky, M. 1985. *The Society of the Mind.* New York: Simon & Schuster.

8. Dutta, S. 1993. *Knowledge Processing and Applied Artificial Intelligence.* Oxford, United Kingdom: Butterworth-Heinemann.

9. Haykin, SO. 2008. *Neural Networks – A Comprehensive Foundation.* Upper Saddle River, New Jersey: Prentice-Hall, 3rd edition, Chapters 1 and 4.

10. McCulloch, WS; Pitts, W. 1943. A logical calculus of ideas immanent in nervous activity. *Bulletin of Mathematical Biophysics* 5. http://deeplearning.cs.cmu.edu/pdfs/McCulloch. and.Pitts.pdf (retrieved 1 January 2017).

11. Haykin, SO. 2008. *Neural Networks – A Comprehensive Foundation.* Upper Saddle River, New Jersey: Prentice-Hall, 3rd edition, Chapters 1 and 4.

12. Ibid.

13. Human Brain Project. https://www.humanbrainproject.eu/ (retrieved 1 January 2017).

14. Wade, N. June 20, 2011. In tiny worm, unlocking secrets of the brain. *New York Times.* http://www.nytimes.com/2011/06/21/science/21brain.html?_r=0 (retrieved 1 January 2017).

15. The Blue Brain Project EPFL. http://bluebrain.epfl.ch/page-56882-en.html (retrieved 1 January 2017).

16. Kurzweil, R. 1990. *The Age of Intelligent Machines.* Cambridge, MA: MIT Press.

17. Bostrom, N. 2014. *Superintelligence: Paths, Dangers, Strategies.* Oxford, United Kingdom: Oxford University Press.

18. Gosling, J; Joy, B; Steele Jr, GL; Bracha, G; Buckley, A. 2014. *The Java Language Specification, Java SE 8 Edition.* Redwood City, CA: Addison-Wesley Professional, 1st edition.

19. Stroustrup, B. 2013. *The C++ Programming Language.* Redwood City, CA: Addison-Wesley Professional, 4th edition.

20. Ibid.

21. Gosling, J; Joy, B; Steele Jr, GL; Bracha, G; Buckley, A. 2014. *The Java Language Specification, Java SE 8 Edition.* Redwood City, CA: Addison-Wesley Professional; 1st edition.

BIBLIOGRAPHY

Human Brain Project. https://www.humanbrainproject.eu/ (retrieved 1 January 2017).

The Blue Brain Project EPFL. http://bluebrain.epfl.ch/page-56882-en.html (retrieved 1 January 2017).

Bostrom, N. 2014. *Superintelligence: Paths, Dangers, Strategies.* Oxford, United Kingdom: Oxford University Press.

Dutta, S. 1993. *Knowledge Processing and Applied Artificial Intelligence.* Oxford, United Kingdom: Butterworth-Heinemann.

Gosling, J; Joy, B; Steele Jr, GL; Bracha, G; Buckley, A. 2014. *The Java Language Specification, Java SE 8 Edition.* Redwood City, CA: Addison-Wesley Professional, 1st edition.

Haykin, SO. 2008. *Neural Networks – A Comprehensive Foundation.* Upper Saddle River, New Jersey: Prentice-Hall, 3rd edition.

Kurzweil, R. 1990. *The Age of Intelligent Machines.* Cambridge, MA: MIT Press.

Lighthill, J. 1973. Artificial intelligence: A general survey. In Artificial Intelligence: A Paper Symposium, Science Research Council. http://www.chilton-computing.org.uk/inf/literature/reports/lighthill_report/contents.htm (retrieved 1 January 2017).

McCarthy, J. April 1960. Recursive functions of symbolic expressions and their computation by machine, Part I. Massachusetts Institute of Technology, Cambridge, MA. http://www-formal.stanford.edu/jmc/recursive.pdf (retrieved 1 January 2017).

McCarthy, J; Minsky, ML; Rochester, N; Shannon, CE. August 31, 1955. A proposal for the Dartmouth Summer Research Project on artificial intelligence. The 1956 Dartmouth Summer Research Project on Artificial Intelligence, Bell Telephone Laboratories. http://www-formal.stanford.edu/jmc/history/dartmouth/dartmouth.html (retrieved 1 January 2017).

McCulloch, WS; Pitts, W. 1943. A logical calculus of ideas immanent in nervous activity. *Bulletin of Mathematical Biophysics*, 5. http://deeplearning.cs.cmu.edu/pdfs/McCulloch.and.Pitts.pdf (retrieved 1 January 2017).

Minsky, M; Papert, SA. 1969. *Perceptrons – An Introduction to Computational Geometry.* Cambridge, MA: MIT Press.

Minsky, M. 1985. *The Society of the Mind.* New York: Simon & Schuster.

Stroustrup, B. 2013. *The C++ Programming Language*. Redwood City, CA: Addison-Wesley Professional, 4th edition.

Turing, AM. October 1950. Computing machinery and intelligence. *Mind* 59 (236). http://www.jstor.org/stable/2251299 (retrieved 1 January 2017).

Wade, N. June 20, 2011. In tiny worm, unlocking secrets of the brain. *New York Times*. http://www.nytimes.com/2011/06/21/science/21brain.html?_r=0 (retrieved 1 January 2017).

Weizenbaum, J. January 1966. ELIZA – A computer program for the study of natural language communication between man and machine. *Magazine Communications of the ACM* 9 (1). http://dl.acm.org/citation.cfm?id=365168 (retrieved 1 January 2017)

CHAPTER 6

What's Next?

The almost exclusive focus on Boolean logic has to some extent hampered the design of computer hardware as we still are reliant on the architecture established at the advent of the first computers. Attempting to develop a computer architecture based on our understanding of the mind provides an interesting dimension; however, the definition of the mind is still somewhat elusive. The mind as the source of human thinking is more than just what the traditional definition of intelligence provides, the concept of rational thinking in accordance with inference and rules of logic. It is well established that man makes decisions based on a view of reality that is constrained through social considerations of what is acceptable and what is not, generally labelled as zeitgeist. Within these realms, applying logic to arrive at rational decisions can be expected, and on that count the more rational one is, the more intelligent. However, there exists a reality outside of the zeitgeist and the perceptions absorbed from it are still registered by the mind but unconsciously so, where they are managed under a very different set of logic. And just because these perceptions are repressed does not mean that our decision-making capabilities are not susceptible to their impact, in fact they are and that too to a considerable degree. Theories on how the unconscious influences human behaviour are century-old but as they have mostly been described in anecdotal fashion, it has reduced their prospects of verifiability and acceptance in other fields of science. But with the progress in neuroscience, we can now ascertain the existence of an unconscious part of the mind. Although not fully explored and documented, it broadly concurs with the psychological theories, namely, it harbours perceptions sorted out as unacceptable by social norms; it consists of drives, previously labelled instincts, which seek to rectify psychologically unsound conventions, not unlike mankind's spontaneous *survival instinct*. It is also proactive in decision making and influences human behaviour in manners often described as irrational. And while such actions might correctly be described as irrational from the perspective of the conscious part of the mind, however, in the holistic context they are sensible expressions based on additional information, *unknown knowns*. It operates on a logic that is association-based and relaxes the mathematical properties of Boolean logic.

The good news is that these metaphysical components can be developed into a model and structured into mechanistic rules without compromising too much and creating a too great discrepancy between a virtual mind and a human mind. The unconscious is governed by a logic and must be as otherwise chaos would reign in our minds, we can see evidence from when these rules have been wrecked in havoc, such as for schizophrenics, fantasy and reality seems indiscriminately intermingled whereas a mentally healthy individual can clearly distinguish between fantasy, with its absence of perspectives of time and traditional cause–effect, and reality. To model a virtual mind, one has two options: either to replicate it on a single individual or on a perceived average mind representing a greater collective; a society, a nation or a cultural domain. The latter is the preferred option for two reasons: the representation of data, both conscious and unconscious expressions, can be drawn as a proxy from public data, which is readily available, and the outlook for the average citizen's mind can be rolled up into the collective, which allows for projections of larger scale changes in the psychological environment that influence the whole society.

But a challenge remains, how does one forecast when unconscious logic with its unknown knowns supersedes the conscious part of the mind with its rationality applied on a truncated reality? A governing protocol needs to be a design that can determine the switches between the different logics. Important to note is that these switches operate on two time horizons: a short-term impact, which is the typical Freudian slip, the intuition, the action in affect, something that is seemingly out of order, *'he was not himself today'* a situation which most are familiar with; and a long-term impact, which probably is a less familiar concept and more important from the societal perspective, a change in zeitgeist that holds the propensity to lead to drastic changes in the cultural, social and political atmosphere.

This book provides a first attempt to model the functionality of the human mind, by breaking down the mind into its verified components, a conscious and an unconscious part with distinctive characteristics. The blueprint presented highlights the design concepts which provide the reader with the opportunity to develop software and hardware solutions and progress artificial intelligence into its next generation. So, while there obviously are computers that are more intelligent, or rather *better*, than humans in certain aspects, including chess computers and the likes, these are still far from able to comprehend and reason like a human. A virtual mind can cover a wide array of applications which current computer architecture finds not feasible, namely, the forecasting and scenario testing of human behaviour, understanding when thought narratives are likely to change, what thought system would be deployed in any given situation. In essence, the whole socioeconomic spectra can be captured, including politics, financial markets and consumer behaviour. Another area of application is to augment various game software and of course, it would be applicable for the man–machine connect as well. The evolving work of these practical applications of a virtual mind by applying it into robots, *psychobots* of sorts, will be described in a coming sequel.

Index